T0213580

Life After Death: What Happens to Your Body After You Die?

Michael Wilson

Life After Death: What Happens to Your Body After You Die?

 Springer

Published in association with
Praxis Publishing
Chichester, UK

Michael Wilson MRSC, MSc, PhD, FRCPath, DSc
Emeritus Professor of Microbiology
University College London
London, UK

SPRINGER-PRAXIS BOOKS IN POPULAR SCIENCE

Popular Science
ISSN 2626-6113   ISSN 2626-6121   (eBook)
Springer Praxis Books
ISBN 978-3-030-83035-9   ISBN 978-3-030-83036-6   (eBook)
https://doi.org/10.1007/978-3-030-83036-6

"Cover artwork by Wilson Harvey, https://www.wilharvey.co.uk/"

This Springer imprint is published by the registered company Springer Nature Switzerland AG
The registered company address is: Gewerbestrasse 11, 6330 Cham, Switzerland

This book is dedicated to Carolyn Cross, an entrepreneur and philanthropist, who has devoted her life to developing a new approach (known as "photodisinfection") for preventing and treating infectious diseases. Her innovative technology has cured many people suffering from such diseases as well as preventing others from catching them in the first place. Thanks to her, therefore, the onset of the processes that are described in this book will be delayed for huge numbers of people.

Preface

There must be very few of us who haven't considered, at some point in our life, the question of what happens after we die. Most of us muse on this in a rather detached, philosophical fashion when we are young, but usually regard it as something that doesn't really affect us and quickly move on to occupy ourselves with more exciting activities. However, as we grow older, we generally get more interested in this question. Once we reach the age of around 70 it finally dawns on us that death is something that really is going to happen to us, probably in the not-too-distant future. Although we start to think more about it, we usually get no nearer to finding an answer to that intriguing "what comes after" question.

Many religions claim to know what happens after we die and this provides comfort to huge numbers of people. But for the rest of us, death remains an interesting, rather worrying, one way journey into the unknown. However, while we may not be able to work out what happens to the essential "me" after death, we can take comfort in knowing what happens to the physical carrier of the "me" i.e. our body. Yes, it will certainly decompose and become smelly. Yes, it will be eaten by microbes and insects and all sorts of other creatures. But, think positive! Our body is a very rich and valuable source of nutrients that will end up feeding huge numbers, and an immense variety, of other creatures. You will end up being recycled in the biosphere and will support the continuation of life on our beautiful planet - this book describes how this comes about.

Chapter 1 is an introduction to the topic that provides an overall perspective on what happens to our body once we die and the benefits this has for other creatures and the environment. The human body is then discussed in terms of it being an important source of a wide range of valuable nutrients.

The third chapter focusses on the concept of a human being as a symbiotic association consisting of a mammal and a variety of microbial communities known collectively as the "human microbiota". This symbiosis is a complex, highly-evolved biological association which benefits both the mammalian and microbial components and this relationship is discussed in detail. Death is followed by a series of physical changes to our body and these are described in the next chapter. While death marks the end of life for their human host, our microbiota lives on, although the composition of the microbial community at a particular body site begins to change. The human microbiota, with the help of environmental microbes, makes sure that the constituents of its human host are recycled so enriching Earth's biosphere and this is the subject of Chapter 5. Insects also play an important role in the decomposition process and, as in the case of microbes, different insects become active as this progresses. The types of insects involved, and the contribution they make to the decomposition process, are described in the succeeding chapter. Finally, the impact that nutrients from a human corpse have on the environment is revealed in the last chapter which also has a little poem to send you on your way, hopefully in a positive frame of mind.

For those of you who would like to know more, a reading list has been included at the end of each chapter. Many of the references provided, whether they are websites or scientific papers, have been chosen because they can be accessed freely by the general public. However, some of the more specialised ones are not freely-available and you'll have to try and obtain them through a library.

Images of dead human beings and other animals can, of course, be very disturbing to look at. Consequently, to spare the feelings of readers, those images that may be upsetting have been included in an Appendix at the end of the book. These can be viewed by those whose curiosity is able to outweigh their sensitivity.

Michael Wilson
University College London
London, UK

Contents

1

It Comes to Us All

So, you've had a long and productive life. You've won the Nobel prize for Literature, you're the proud parent of three healthy and successful children who have provided you with six adorable grandchildren. You are a pillar of your local community and have contributed greatly to the wellbeing of young and old through your charitable works. You've a lovely house, a cottage in the country and a solid investment portfolio. But within the next 60 seconds you're going to be dead (Figure 1.1) and then, as Macbeth says (Act V, Scene 5) on learning of the death of his wife:

"And all our yesterdays have lighted fools
The way to dusty death. Out, out, brief candle!
Life's but a walking shadow, a poor player
That struts and frets his hour upon the stage
And then is heard no more: it is a tale
Told by an idiot, full of sound and fury,
Signifying nothing."

So, what about you? Will you "go gentle into that good night"[1] or will you "Rage, rage against the dying of the light" (see Footnote 1). Who knows? (Box 1.1). You don't have some dreadful disease and won't have a painful death, thanks to a range of painkillers provided by the pharmaceutical industry and administered by your attentive doctor.

---

[1] "Do Not Go Gentle Into That Good Night" by Dylan Thomas, from THE POEMS OF DYLAN THOMAS, copyright ©1952 by Dylan Thomas. Reprinted by permission of New Directions Publishing Corp. within the USA, and by permission of The Dylan Thomas Trust outside the USA.

© The Author(s), under exclusive license to Springer Nature Switzerland AG 2022
M. Wilson, *Life After Death: What Happens to Your Body After You Die?*,
Springer Praxis Books, https://doi.org/10.1007/978-3-030-83036-6_1

Figure 1.1   The Death of King Arthur (James Archer c. 1860 CE)
James Archer (1823–1904), Public domain, via Wikimedia Commons

Box 1.1 Death and the poets

Many poems have been written about death and Shakespeare said a lot about it
(he would, of course) but one I particularly like is something I learnt at school
many years ago because it offers a defiant challenge to Death, a great
confrontation:

> Death, be not proud, though some have called thee
> Mighty and dreadful, for thou art not so;
> For those whom thou think'st thou dost overthrow
> Die not, poor Death, nor yet canst thou kill me.
> From rest and sleep, which but thy pictures be,
> Much pleasure; then from thee much more must flow,
> And soonest our best men with thee do go,
> Rest of their bones, and soul's delivery.
> Thou art slave to fate, chance, kings, and desperate men,
> And dost with poison, war, and sickness dwell,
> And poppy or charms can make us sleep as well
> And better than thy stroke; why swell'st thou then?
> One short sleep past, we wake eternally
> And death shall be no more; Death, thou shalt die.

This poem is called "Death be not proud" and was written by John Donne
(1572-1631) who, as well as being a poet, was Dean of St. Pauls Cathedral
in London.

Box 1.1 (continued)

Bust of John Donne with St Paul's Cathedral in the background
Archaeomoonwalker, CC0, via Wikimedia Commons

But what next? It'll be the end of many tender experiences including those mentioned in Lucinda Williams' beautiful song "Sweet Old World" https://www.youtube.com/watch?v=G3DEjpSekNY

You'll be greatly missed by your life-long partner, the rest of your family and by your extensive network of friends. They'll cherish memories of you for many a year and will relate treasured stories of events in your life. But what about your body? What's going to happen to that? Well, as the bible tells us, it's basically a case of "dust thou art, and unto dust shalt thou return". But you shouldn't let that depress you, you'll be going on a fascinating and extremely complex journey and the purpose of this book is to describe this for you and explain what it will involve. When push comes to shove, once you die, all you are is a complex mixture of chemicals enclosed within a bag of skin. You are now a corpse, a word derived from the Latin word "corpus" which means "a body" via the French word "corps". Your longings, ambitions and achievements may have all vanished but you're much more than the sum of those – you are also an important source of nutrients for a huge variety of creatures, both microscopic and macroscopic. Rejoice! You have a new purpose – you are about to be recycled. You will undergo a process that is fundamental to, and characteristic of, all the natural materials on our planet. This is because you share an important property with such materials - you are biodegradable. In other words you can be broken down and re-used by other organisms. You share this characteristic with all the other creatures, large or small, that live on our planet. The only things that can't be recycled in this way are many of the substances made by humans such as plastics (Box 1.2).

Box 1.2 Biodegradation and food webs

What do we mean when we say that something is biodegradable? Literally, this word means "can be broken down to simpler molecules by living organisms". The term is usually used with respect to large molecules or polymers (described in greater detail in Chapter 2). The organisms mainly responsible are microbes such as bacteria and fungi (you'll learn much more about microbes in Chapter 3) and insects (see Chapter 6). All living creatures on our planet, as well as the substances they produce, are biodegradable. This means that nothing is ever wasted and all the materials that make up, or are produced by, living organisms are eventually re-used to make more organisms.

A simple example of the biodegradation and recycling process, with which most of us will be familiar, is the cow-dung-plant system (Figure a). Such a system is known as a "food web" as it involves feeding interactions between different organisms. A cow eats grass and produces dung which falls to the ground. The dung is a complex mixture of large and small molecules. The small molecules are absorbed into the soil. The large molecules in it (such as cellulose) are broken down by insects and microbes resulting in small molecules and ions (molecules that have a positive or negative charge) which are absorbed into the soil. All of the molecules and ions (such as phosphate, nitrate, sulphate, iron, potassium and calcium) are then used by grass and other plants to grow and reproduce which results in more plants. The plants are then eaten by the cows and the whole process is repeated. Another feature of this system is that the insect larvae that feed on the dung develop into adults and these are involved in the pollination of flowering plants.

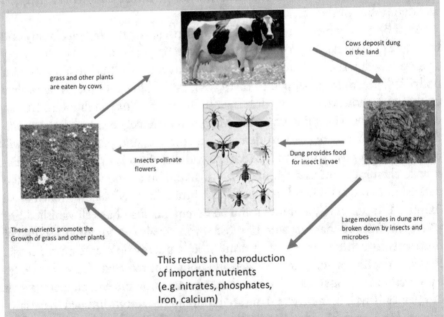

Figure (a) The cow-dung-plant system

Box 1.2 (continued)

<u>Image credits</u>
Cow: Keith Weller/USDA, Public domain, via Wikimedia Commons
Dung: Wilfredor, CC0, via Wikimedia Commons
Insects: Robert Evans Snodgrass, Public domain, via Wikimedia Commons
Grass/flowers; author

Unfortunately human ingenuity has resulted in the production of materials, such as plastics, that aren't biodegradable. Most types of plastic can't be broken down by microbes or insects. They can, therefore, survive in the environment for many years and, as well as being unsightly, can't act as nutrients for other creatures (Figure b). The problems of the persistence of plastics in our soils and oceans are well known and are a major threat to life on our planet.

Figure (b) Non-biodegradable plastic rubbish that will persist in the environment for hundreds of years.

Once you were a provider of love, encouragement, support, food and material resources. But, even after you're dead, you'll continue to be a provider – of nutrients for a range of creatures you would probably have ignored while you were alive. This is amazing. The chemicals of which you are composed are going to be re-used and circulated throughout the environment for the benefit of countless organisms. Your atoms and molecules will be disseminated huge distances and incorporated into the cells of many different types of microbes, plants and animals. In accordance with the laws of Wave Mechanics you will permeate the universe and you will truly become "a child of the universe" (Max Ehrmann, *"Desiderata: A Poem for a Way of Life"*).

This fascinating journey will start with your burial. Although cremation is becoming increasingly popular, large proportions of the world's population still end up in graves rather than being cremated. In the Islamic and Jewish religions, burial is preferred over cremation and this also applies to the Roman Catholic and Orthodox churches. In contrast, Hindus and Sikhs are usually cremated. Buddhists tend to follow whatever practice is prevalent in the country in which they live. Cremation results in the production of ashes (Box 1.3) which contain a high proportion of phosphates which, of course, will enrich the soil if these are buried. The process converts the organic matter present in our tissues to carbon dioxide and each cremated body produces approximately 243 kilograms of CO_2. From an ecological point of view, cremation is a bad idea because it not only produces lots of CO_2 but it also deprives a wide range of microbes, animals and plants of the nutrients they need for their growth and proliferation. Wouldn't you much rather feed these multitudes directly?

Box 1.3 Cremation

Cremation involves heating a corpse to a temperature (800 to 1000°C) that is high enough to make it burn and this results in the production of human ashes (often referred to as "cremains") after 2-3 hours. During this heating process the water present in your body is first driven off, the soft tissues burn and bones become calcified. At the end of the process all that remains are ashes and bone fragments. These are then placed into a grinding machine (known as a cremulator) to break up the bone fragments and the result is a sandy, greyish material. The cremains of a typical adult male and female weigh approximately 6 and 4 pounds respectively. The material is a complex mixture of mainly calcium phosphate and sulphate as well as the salts of a wide range of other metals that were originally present in your body (Figure).

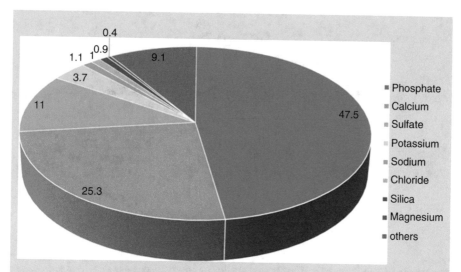

Figure. Approximate composition (% by weight) of human cremains. "Others" includes a variety of metal ions such as Aluminium, Iron, Zinc, Copper, Manganese etc.

Despite its high content of phosphate and other minerals important for plant nutrition, the addition of cremains to soil doesn't necessarily result in an increase in its fertility. This is because the high temperature used to cremate the body results in the production of an insoluble form of calcium phosphate which reduces its availability to plants. Furthermore, the high content of sodium ions is toxic to plants. However, cremains can be treated with acid and/or mixed with compost before being added to soil and this results in the release of a more soluble form of phosphate that plants can use.

The oldest evidence of a human cremation found so far comes from remains found at a burial site near Lake Mungo in New South Wales, Australia. These are of a female (known as the "Mungo Lady") and are estimated to be about 40,000 years old. The body was cremated, the remains were then crushed and then burnt again before being buried. In Eurasia, cremation appears to have been practiced much later as the oldest example found so far is that of the 9000 year old cremains of a male found at a burial site in the Upper Jordan Valley, Israel.

In ecological terms, your body is a very valuable resource because of the large variety and quantity of nutrients it contains, as will be described in Chapter 2. Microbes, insects, animals and plants can all make use of the nutrients present in a human corpse (Figure 1.2).

Let's look at Figure 1.2. more closely. The figure shows that a human corpse can be used by insects, animals other than insects and microbes. However, in the case of most corpses, animals are rarely involved because most are buried (Box 1.4) rather than left where they can be found by wild or domesticated animals.

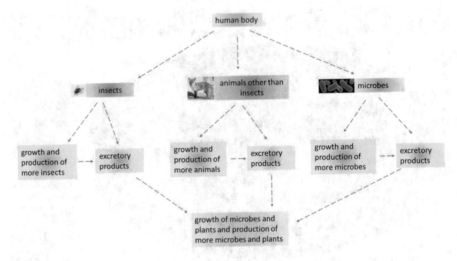

Figure 1.2 Utilisation of nutrients present in a human corpse
Fly: USDAgov/Public domain
Fox: Dan Davison from Rochford, England/CC BY (https://creativecommons.org/licenses/by/2.0)
Bacteria: Jennifer Oosthuizen, Centers for Disease Control and Prevention, USA

Box 1.4. Human burial practices

Burying the dead in the ground is a procedure that has been practiced by humans for at least 100,000 years. The oldest burial site found so far is in the floor of the Qafzeh cave near Nazareth in Israel. The bodies of a group of 15 adults and children were buried in these graves during the Middle Palaeolithic era. There is considerable evidence that corpses were buried in the Mesopotamian, Ancient Greek and Ancient Egyptian cultures. The oldest grave discovered in Europe is that of the Red Lady of Paviland in Wales. In this grave the skeleton of what was thought to be a female, but later shown to be a male, was found. The skeleton had been dyed in red ochre and is of a corpse buried about 33,000 years ago.

Perhaps the oldest grave in North America is that found on the banks of the Columbia River near the town of Kennewick, Washington. It contained the skeleton of a 40-55 year old male (who became known as "Kennewick Man") whose burial was dated to approximately 7,000 BCE.

Nevertheless, occasionally a human corpse is used as a food source by large animals and this practice is known as "scavenging" (Figure 1.3 – shown in Appendix III). This may happen as a result of some unfortunate event such as (i) dying in some remote region, (ii) being murdered and having your body dumped in a rural environment and (iii) dying (or being murdered) at home if you live alone. Scavenging results in dismemberment of the body and, often, scattering of the body parts over a wide area because the animal may take body parts back to its home to feed its young.

An unburied human corpse left outdoors is most likely to be scavenged by large carnivores (wolves, hyenas, jackals, foxes, coyotes, bears, dogs etc.) or by various birds including vultures, crows, magpies, buzzards and ravens. In contrast, the scavengers of corpses left indoors are more likely to be domestic dogs or cats, rats and mice. The particular scavenger(s) involved in the disposal of a human corpse can often be determined because the ways in which they feed on the corpse leave behind characteristic marks on the flesh and the bones.

However, as mentioned previously, because of modern burial practices, most of us will not be eaten by wild animals and our body parts won't be scattered about. Instead, we'll either be cremated or our bodies will be buried where they'll undergo a process known as decomposition i.e. they will be consumed mainly by microbes and insects.

Regardless of the size of the creatures that feed on our corpse, what is happening is, essentially, the breakdown of our tissues (see Chapter 2) to their component molecules and the subsequent use of these as nutrients (Chapters 5 and 6).

The first stage in this process is self-inflicted and doesn't involve any other creatures - the cells of our body start to fall to pieces. This process is known as autolysis and will be described in Chapter 5. What happens next is rather peculiar. Did you know that most of the cells in your body aren't human cells but are microbial cells? Well, this is a fact. Some scientists have suggested that these microbes outnumber our own cells by a factor of 10. But, don't panic. These microbes are good for us. In fact, as you'll find out in Chapter 3, they're essential to our well-being while we're alive. But here is the bad news. Once we die these life-long friends of ours turn around and start to eat us – traitors! How they do this is described in Chapter 5. Later on, they're helped by microbes present in the surrounding soil. Interestingly, a succession of different microbial communities develops with time and identifying which microbes are present may, ultimately, help in forensic investigation of a corpse to establish the time of death.

But microbes aren't the only creatures that feed on our corpse, insects are also attracted to this rich source of nutrients and join in the feast (described in Chapter 6). A number of different insects will lay their eggs in your decaying corpse and it's the larvae that hatch from these that will make most use of the food on offer. As in the case of microbes, the type of insect found in a corpse varies with the time since death and this can aid in establishing how long a person has been dead.

Consumption of the nutrients in our corpse takes place over quite a long time period (a month or two for flesh, but years in the case of our bones) and our appearance changes dramatically during this time. These changes are described in Chapter 4 which, to protect those who are squeamish (which includes the author), will be largely free of images – these are very gruesome. Eventually, all that will be left of our bodies will be bones and teeth. But during this slow process of decomposition the nutrients in our body will have been used by a huge variety of microbes and animals. Furthermore, we'll also have enriched the soil surrounding our grave and so provided nutrients to enable the growth of a variety of plants, microbes and animals. All in all, we'll have made a significant contribution to biodiversity (Box 1.5) and this will be described in Chapter 7.

Box 1.5 Biodiversity

The biodiversity (i.e. biological diversity) of an area (no matter how small or large) is the number of different types of creatures that live there. Because of the many interactions that occur among different creatures (see, for example, Box 1.2) a high biodiversity enables mutual support and so stabilises all life in that particular area. An area with a high biodiversity can adapt to, and resist, the possible adverse effects of change. If the biodiversity of an area decreases then it becomes more vulnerable to any change that occurs there and dramatic changes can eventually lead to the extinction of particular species.

Human activity often results in environmental problems and a reduction in biodiversity. However, the gradual decomposition of your corpse will result in an increase in the biodiversity of the area surrounding your grave.

Exhibition on biodiversity at the Museum of Natural History, Berlin
Anagoria, CC BY 3.0 <https://creativecommons.org/licenses/by/3.0>, via Wikimedia Commons

Do you feel better about dying now? Are you pleased to learn that, even after death, you can contribute to the health of our planet? I hope so. This introductory chapter has sketched out an outline of what will happen to your body after you die. The remaining chapters will, in a reversal of the decay process, put flesh on the bones of this bare skeleton.

1.1   Want to know more?

The corpse project. Can we lay bodies to rest so that they help the living and the earth?
http://www.thecorpseproject.net/

What happens to our bodies after we die. BBC
https://www.bbc.com/future/article/20150508-what-happens-after-we-die

Biodiversity. Natural History Museum, London
https://www.nhm.ac.uk/discover/what-is-biodiversity.html

Biodiversity. Openstax
https://openstax.org/books/concepts-biology/pages/21-1-importance-of-biodiversity

Biodiversity. BBC Bitesize
https://www.bbc.co.uk/bitesize/guides/zs8wwmn/revision/1

Biodiversity. Stanford Encyclopedia of Philosophy
https://plato.stanford.edu/entries/biodiversity/

Food chains and food webs. ThoughtCo
https://www.thoughtco.com/what-is-a-food-web-definition-types-and-examples-4796577

Food chains and food webs. National Geographic Society
https://www.nationalgeographic.org/encyclopedia/food-web/

Food chains and food webs. BBC Bitesize
https://www.bbc.co.uk/bitesize/guides/zq4wjxs/revision/1

Food chains and food webs. K8 School
https://k8schoollessons.com/food-chains-food-webs/

Food chains and food webs. Khan Academy
https://www.khanacademy.org/science/ap-biology/ecology-ap/energy-flow-through-ecosystems/a/food-chains-food-webs

Cremation. How Stuff Works
https://science.howstuffworks.com/cremation1.htm

Cremation. Popular mechanics
https://www.popularmechanics.com/science/health/a18923323/cremation/

Cremation Institute
https://cremationinstitute.com/cremation-process/

Cremation. Intechopen
https://www.intechopen.com/books/biodegradation-life-of-science/
biodegradation-involved-microorganisms-and-genetically-engineered-micro-
organisms

Burial practices. New World Encyclopaedia
https://www.newworldencyclopedia.org/entry/Burial

Burial practices.
https://www.joincake.com/blog/why-do-we-bury-the-dead/

Biodegradation.
http://www.polimernet.com/Docs/Aerobic%20-%20Anaerobic%20
Biodegredation%20en.pdf

Scavengers. National Geographic Society
https://www.nationalgeographic.org/encyclopedia/scavenger/

Scavengers. AnimalWised
https://www.animalwised.com/what-are-scavenger-animals-2774.html

Death Poems and Poetry
http://famouspoetsandpoems.com/thematic_poems/death_poems.html

Best Poems About Death by Famous Poets
https://www.familyfriendpoems.com/poems/famous/death/

Best Famous Death Poems
https://www.poetrysoup.com/famous/poems/best/death

2

A Rich Bag of Goodies - The Human Body as a Source of Nutrients

The earliest traumatic event in my life I can remember occurred when I over-heard my sisters ask my mother what they were made of. They were told "Sugar and spice, and all things nice, that's what little girls are made of". Eager to be reassured, I asked if I was the same. My mother frowned, shook her head sadly, answered "No" and added "Rats and snails and puppy-dogs' tails, that's what little boys are made of." Oh, the horror. But, after many years of therapy and a good grounding in biochemistry, I've managed to get over it. Hopefully, in these "woke" days other little boys won't have to suffer from such sexist unfairness.

A more pleasant memory is being told "You're such a treasure". How many times did your parents and relatives say that to you when you were a child? I must admit that I'm not called that very often nowadays. But, you are indeed a treasure. You are a veritable treasure-trove of nutritious substances that can act as food for an enormous variety of creatures, both large and small. But, before we look at this in more detail, it's important to think for a moment about how your body is organised as this will give you a better understanding of how it works and how it acts as a source of nutrients.

Your body is organised on several levels (Figure 2.1). First of all, it can be viewed as consisting of a number of systems each of which has a particular function such as digestion, reproduction and respiration. Each of these sys-tems is made up of a set of organs. For example, the digestive system consists of the stomach, liver, large intestine etc. Each organ is itself made of tissues; a tissue being a group of cells and other materials that work together to perform a particular function.

M. Wilson, *Life After Death: What Happens to Your Body After You Die?*, Springer Praxis Books, https://doi.org/10.1007/978-3-030-83036-6_2

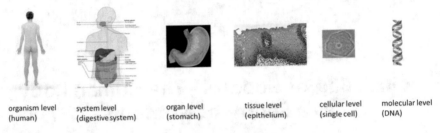

organism level   system level   organ level   tissue level   cellular level   molecular level
(human)   (digestive system) (stomach)   (epithelium)   (single cell)   (DNA)

Figure 2.1 The various levels of organisation in a human
Human male: Image uploaded by Mikael Häggström, CC0, via Wikimedia Commons
Digestive system: OpenStax College, CC BY 3.0 <https://creativecommons.org/licenses/by/3.0>, via Wikimedia Commons
Stomach: Mikael Häggström, Public domain, via Wikimedia Commons
Epithelium: Berkshire Community College Bioscience Image Library, CC0, via Wikimedia Commons
Cell: M. Rein, Centres for Disease Control and Prevention, USA
DNA: Data Base Center for Life Science (DBCLS), CC BY 4.0 <https://creativecommons.org/licenses/by/4.0>, via Wikimedia Commons

The organisms that will eventually feed on your corpse aren't, of course, interested in the higher levels of organisation within your body – all they're really after is human tissue because this is what can provide them with nutrients. Our focus in this chapter, therefore, is going to be the nutrient content of your tissues. But first of all, a word of warning. This chapter is going to venture into the realms of biochemistry – the very chemistry of life. This is a very exciting world to explore but it will mean you're going to come across a lot of new concepts. So, be prepared to be exercise those brain cells while you can – these, of course, will provide a tasty meal for some microbes and insects in the future (Dreadful joke warning #1 – this is a case of "thought for food" rather than "food for thought"). The chapter also contains details of human anatomy, but this is likely to be more familiar to you than the biochemistry.

2.1   What Types of Tissues Are Present in My Body?

There are four different types of tissue – epithelial, connective, muscle and nervous.

2.1.1   Epithelial Tissue – Keeping It All Together

Epithelial tissue (Figure 2.2) covers your body's surfaces and consists of tightly-packed epithelial cells which is an arrangement designed to protect and defend your body from the external environment.

The epithelial tissue that covers your external surfaces is known as skin (Figure 2.3). This, basically, delineates you from the outside world – whatever lies within your epithelial tissue is "me". Beyond your skin is the external environment, the outside world, the "not me" part of the rest of the vast universe, with all of its potential dangers. The skin protects you from this external world but also allows you to communicate with it – it does this by means of sensors (for touch, temperature, taste etc.) that are embedded in the skin.

The epithelium that lines your body cavities, such as the respiratory, gastrointestinal and genitourinary tracts, is different. It's known as a mucosa (or mucosal surface or mucus membrane) because it's covered in a layer of mucus (Figure 2.4) and is therefore moist, unlike most of the skin which is normally quite dry. Mucus consists mainly of water but also contains mucins which make it sticky. Mucins are very large molecules - biochemists call these "macromolecules". Other macromolecules we'll be talking about in this chapter include proteins, glycoproteins, proteoglycans, nucleic acids, polysaccharides and lipids.

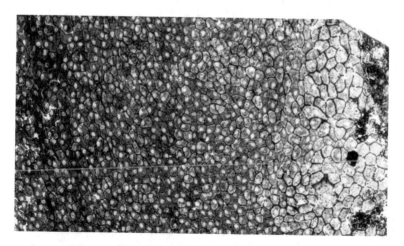

Figure 2.2   Human epithelial cells showing their tightly-packed arrangement (x200). The individual cells can be seen to form a tightly-packed mosaic with no spaces between them. The nucleus in each cell appears as a lighter-coloured oval shape
Image courtesy of Berkshire Community College Bioscience Image Library, Public Domain CC0 1.0 Universal (CC0 1.0)

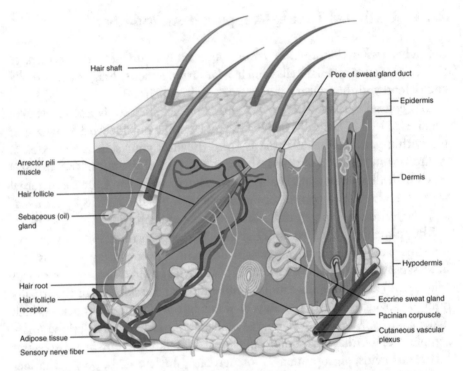

Hair shaft

Pore of sweat gland duct

Epidermis

Arrector pili muscle

Hair follicle

Sebaceous (oil) gland

Dermis

Hair root

Hair follicle receptor

Adipose tissue

Sensory nerve fiber

Hypodermis

Eccrine sweat gland

Pacinian corpuscle

Cutaneous vascular plexus

Figure 2.3 A diagram of a cross-section through human skin showing that it consists of three main layers – epidermis, dermis and hypodermis. The epidermis itself consists of a number of layers and the outermost of these (known as the stratum corneum) is made up of tightly-packed dead cells held together by fats (biochemists call these "lipids"). This is the layer that is in contact with the external environment. Hairs arise from follicles and their orientation is controlled by muscles (arrector pili muscles). Sweat is produced by eccrine (sweat) glands and the sebaceous glands produce an oily substance known as sebum

You become very aware of mucus when you have a cold – that's when nasal mucus (snot) starts to pour from your nose. This is a great defence mechanism – your body is trying to get rid of the virus that's causing your cold by flushing it out from the upper regions of your respiratory system.

One of the most important functions of both types of epithelial tissue is to stop microbes from the environment getting access to our sterile internal tissues and organs. In Chapter 4 we'll be able to see the dramatic consequences of the breakdown of these barriers after we die.

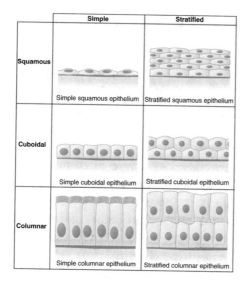

(a)

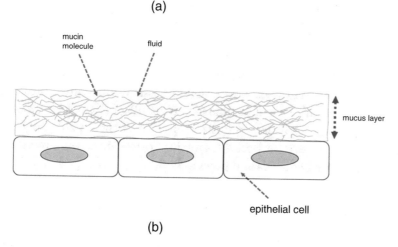

(b)

Figure 2.4 The epithelium that lines body cavities such as the mouth, lungs, throat and gut is known as a mucosa and is covered in a layer of a sticky fluid known as mucus (a) Mucosal surfaces consist of either a single layer of cells ("simple") or several layers ("stratified"). The shape of the cells also differs in different tissues, depending on the function of the particular tissue
OpenStax College/CC BY (https://creativecommons.org/licenses/by/3.0)
(b) The layer of epithelial cells is covered with a layer of a sticky fluid known as mucus. Mucus consists mainly of water and long molecules of mucin which is a glycoprotein

2.1.2   Connective Tissue – Our Support System

Connective tissue (Figure 2.5) is the most abundant, and widely distributed, of the four tissue types and is involved in binding together, supporting and strengthening other tissues. It's composed of cells that are embedded in a gel-like supporting structure known as the "extracellular matrix". This matrix is a mixture of various macromolecules and is permeated by fibres that strengthen it and bind it together.

Blood (Figure 2.5b) is also classed as a type of connective tissue (even though it's a liquid) and is the main way in which things are moved around the body i.e. it's the body's transport system. It transports nutrients, oxygen and components of our immune system (antibodies and white blood cells) that help to defend us against microbes. Fat is stored in a special type of connective tissue known as adipose tissue (Figure 2.5c). This is an important energy reserve and also protects and insulates the body. Finally, bone is a very hard type of connective tissue that's involved in supporting us, protecting our vital organs and enabling us to move. There are lots of different types of bone (Figure 2.5d), depending on their particular function, but all are composed of two main materials – hydroxyapatite (about 70%) and organic polymers (about 30%). Bone cells are also present and these are of two main types - osteoblasts (which synthesise new bone) and osteoclasts (which break down bone). Hydroxyapatite is a mineral made up mainly of calcium phosphate and calcium carbonate. Collagen comprises 90–95% of the organic polymers of bone, the rest consists of various proteins (osteocalcin, osteopontin, and osteonectin) as well as lipids and polysaccharides.

2.1.3   Muscle Tissue – Standing and Moving

Muscle tissue (Figure 2.6) enables us to move as well as maintain our posture and also generates heat – it comprises about 40% of the total mass of the body of an adult. The meat eaten by those of us who aren't vegetarians consists

Figure 2.5 (continued)   Image courtesy of the Berkshire Community College Bioscience Image Library. CC0 1.0
(c) Adipose tissue (diagram on the left and photomicrograph on the right). This is a tissue involved in the storage of fats (i.e. lipids) which are an important energy reserve. It consists of cells known as adipocytes which contain lipids in the form of large droplets.
OpenStax College/CC BY (https://creativecommons.org/licenses/by/3.0)
(d) the various types of bone that make up the human skeleton
BruceBlaus, CC BY 3.0 <https://creativecommons.org/licenses/by/3.0>, via Wikimedia Commons

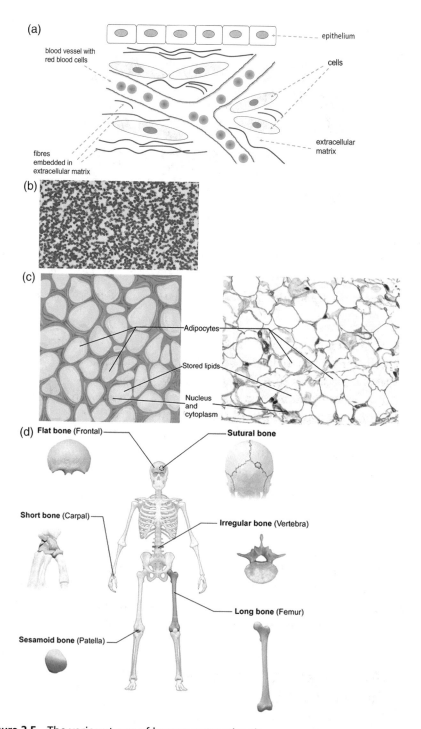

Figure 2.5 The various types of human connective tissue

(a) Diagram showing the main features of connective tissue. It consists of cells embedded in an extracellular matrix consisting of macromolecules and usually contains fibres.
(b) Photomicrograph of blood showing mainly red blood cells. A few different types of white blood cells (stained blue) can also be seen

Figure 2.6 Photomicrograph of skeletal muscle (X100). Long fibres, known as myofilaments, (stained red) can be clearly seen – these are able to contract and relax to enable movement. They are composed of the proteins myosin and actin. The nuclei of cells (stained blue) are also visible
Image courtesy of the Berkshire Community College Bioscience Image Library. CC0 1.0

mainly of muscle. It's made up of cells and two proteins – myosin and actin. Muscle cells form fibres that may be arranged in bundles (in skeletal muscle) or in sheets (smooth muscle) that surround many organs. Skeletal muscle is attached to bones by means of tendons and enables us to move. The muscles that surround many organs propel material through those organs. For example, smooth muscles are used to propel blood through the heart and digested food along the intestinal tract.

2.1.4   Nervous Tissue – "I've Got a Feeling" (Thanks to The Beatles)

Nervous tissue (Figure 2.7) is not a name given to tissue that's very sensitive and frightened of everything (dreadful joke #2). It's a type of tissue that's involved in coordinating the body's activities by responding to stimuli and transmitting nerve impulses. These nerve impulses can make muscles contract as well as stimulate organs and glands to respond in some way.

2.2   But What Are All These Tissues Made Up Of?

What's in a tissue? Well, aerosols and droplets of snot for a start (dreadful joke #3). The human body consists mainly of six elements – oxygen, carbon, hydrogen, nitrogen, calcium and phosphorus (Figure 2.8a), although there

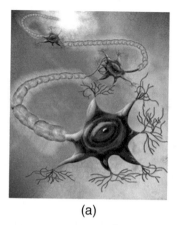

(a)

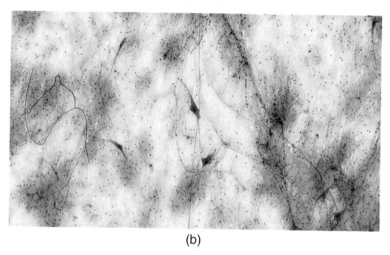

(b)

Figure 2.7   Nervous Tissue

(a) Diagram of three nerve cells joined together. A neuron (nerve cell) is an oddly shaped cell that has a large star-shaped central "cell body" containing the nucleus. The cell body has many thin branches (dendrites) which receive incoming signals from other nerve cells via junctions known as synapses. The cell body also has a long, thin projection called an axon (protected by a waxy sheath – coloured orange) which ends in several thin terminals which connect with another neuron, a muscle or a gland. A signal is received by the dendrites, passes along the axon and is then transmitted to another neuron (or a muscle or gland) via the axon terminals

Bill McConkey. Attribution 4.0 International (CC BY 4.0)

(b) Photomicrograph showing nerve cells in the spinal cord. The cell bodies of several nerves can be seen as well as the long axons

Berkshire Community College Bioscience Image Library/CC0

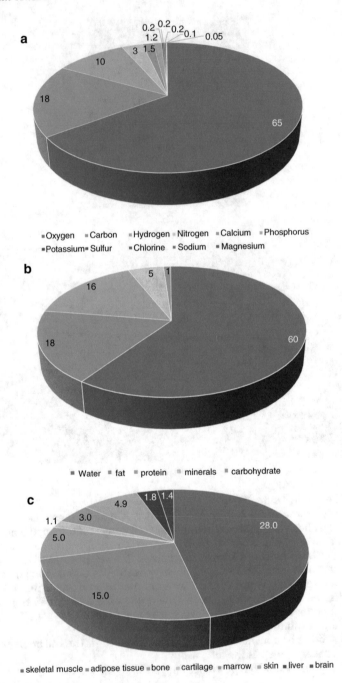

Figure 2.8 Composition of an adult human.
(a) relative proportions (percentages by weight) of the main elements.
(b) Relative proportions (percentages by weight) of the main types of compounds present.
(c) Weight (in kilograms) of major tissues and organs present in an average body weighing 70 kilogrammes

are smaller quantities of a wide range of other elements. But, a more important question for the creatures that will eventually feed on our bodies, is "in what types of compounds are these elements present in our body"? The total weight of tissues in the average adult is about 70 kg and most of this is water – around 62% (Figure 2.8b). A fresh corpse, before it dries out, therefore constitutes an excellent supply of water for other creatures. The other main components are fat, protein and carbohydrates (Box 2.1) - these being the building blocks of the cells and the other materials that make up our tissues. A variety of nucleic acids, small molecules and minerals are also present, but in much smaller quantities.

Box 2.1 Carbohydrates, sugars, saccharides, oligosaccharides and polysaccharides – what's in a name?

A carbohydrate is, strictly speaking, a molecule that contains the elements carbon, hydrogen and oxygen with the ratio of hydrogen (H) to oxygen (O) being 2:1. As the H:O ratio is 2:1 (i.e. the same as in water) the molecule can be considered to be a "hydrate" of carbon – hence the name. The word "saccharide" (from the Greek word sakkharon, which means sugar) is simply another name for a carbohydrate and the terms are interchangeable.

The structure of the simplest carbohydrates consists of a ring of carbon atoms (often 5, but this can be more or fewer) and an oxygen atom with other atoms, particularly hydrogen and oxygen atoms, attached to this (Figure).

Figure. Diagram showing the structure of glucose which is a monosaccharide (or simple sugar).

The glucose molecule shown above is an example of what is known as a monosaccharide because it consists of a single saccharide unit. These units can link together to form larger molecules consisting of two, three or even thousands of sugar units. They are all carbohydrates but, confusingly, other names are used

Box 2.1 (continued)

for them in addition. A molecule that consists of a small number of monosaccharide units is known as a disaccharide, trisaccharide etc. depending on the number of units present. If the number of monosaccharide units is between 3 and 9 it's then called an oligosaccharide. When it reaches more than 9 it's called a polysaccharide. The term sugar is, confusingly, also used to refer to carbohydrates that contain less than 3 monosaccharide units.

The basic sugar molecule may also have other chemical groups attached to the ring such as amino and sulphate groups.

The sugars that human beings crave (and which do so much damage to our teeth and help us to put on the pounds) are mainly simple sugars such as sucrose (a disaccharide), glucose (a monosaccharide) and fructose (a monosaccharide).

The number of different types of proteins, fats, carbohydrates and minerals that comprise human tissues is huge. In addition, there are smaller quantities of other nutritionally-important compounds such as nucleic acids and vitamins. The human body, therefore, contains a wide range of compounds that are potential nutrients as well as water which, of course, is an important resource for all living organisms.

2.2.1   Small Is Beautiful, But Big Is Also Very Important

Many of the constituents of human tissues are compounds with a low molecular mass (water, vitamins, organic acids etc.) or ions (sodium, chloride etc.) that can be used directly as nutrients by microbes and other organisms. However, more than one third of the tissue mass (or, if we ignore water, 86% of the dry mass) is in the form of compounds with a large molecular mass – these are known as macromolecules and include proteins, glycoproteins, proteoglycans, lipids, polysaccharides and nucleic acids. Such large molecules must first of all be broken down into smaller compounds before they can be used by microbes or insect larvae which, as mentioned in Chapter 1, are the main organisms responsible for the decomposition of a human corpse.

We humans are, of course, faced with the same problem when we eat. Our food (whether animal or plant) also consists of these two chemical types of potential nutrients – small and large molecules. The small molecules and ions (e.g. simple sugars, amino acids, sodium ions, chloride ions) that are present in our food don't need any further processing and are absorbed directly into our bloodstream in the stomach and small intestine. These are immediate sources of the energy needed to power our body and of the materials that can be used to make new body cells. However, the macromolecules in our food (e.g. proteins, polysaccharides, nucleic acids, lipids) need to be broken down into smaller molecules before they can be absorbed into our bloodstream and used as nutrients. This breakdown occurs in the mouth, stomach, small intestine and large intestine. It's carried out by enzymes which are present in the mouth (in saliva), the stomach (in gastric juice) and in the other regions of the intestinal system.

In Figure 2.8 we can see that most of the potential nutrients present in a human corpse are in the form of macromolecules (mainly proteins and fats) so we need to learn a bit more about these, and other, macromolecules. Now we're really going to have to delve into the depths of biochemistry.

2.2.2   Structures of the Main Macromolecules in Humans

How do we decide whether a molecule is classified as being a macromolecule or not? Well, it's all matter of weight or, strictly speaking, mass. The mass of an atom, molecule or ion (i.e. a charged atom or molecule) used to be expressed in terms of how much heavier it was than a Hydrogen atom – this being the simplest atom there is. The Hydrogen atom was considered to have a mass of 1. Atomic, molecular or ionic masses were, therefore, relative values rather than absolute values. However, chemists in 1961 decided that this wasn't good enough (far too simple) so they decided to compare the masses of atoms, molecules and ions to that of an atom of Carbon, instead of Hydrogen. A carbon atom was allocated a mass of 12 Daltons (Da). The molecular masses of some important small molecules and ions found in human tissue are 18 Da for water, 23 Da for the sodium ion, 60 Da for urea, 146 Da for lysine (an amino acid) and 180 Da for glucose (a sugar). Macromolecules, in comparison, are huge and can contain many thousands of atoms and so have very large molecular masses. For example the molecular mass of collagen (an important protein in connective tissue) is about 300,000 Da while that of hyaluronan (a polysaccharide in connective tissue) is 8,000,000 Da. Most of the major macromolecules in human tissues are polymers i.e. they're made from small molecules (known as monomers, repeating units or building blocks) that have been joined together like the beads on a necklace (Table 2.1 and Figures 2.9-2.11).

Table 2.1 The main polymeric macromolecules found in the human body

Macromolecule	Monomer (repeating unit or building block)
protein	amino acid
polysaccharide	sugar
nucleic acid	nucleotide (a complex molecule consisting of a sugar, a base (adenine, thymine, guanine or cytosine) and phosphoric acid

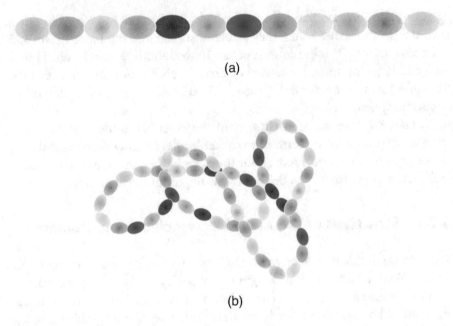

Figure 2.9 Protein structure
(a) Diagram showing the primary structure of a protein. Each of the oval shapes is an amino acid and these are all joined together.
(b) The long chain of amino acids then folds to produce a complex shape that is characteristic of that particular protein.

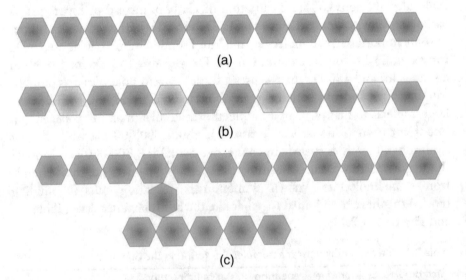

Figure 2.10 Polysaccharides consist of monosaccharide units joined together. They may form a linear arrangement (a and b) which all contain either the same monosaccharide (a) or contain different monosaccharides (b). Alternatively, they can be arranged to form a branched structure (c) that may consist of one type of monosaccharide or different types of monosaccharides

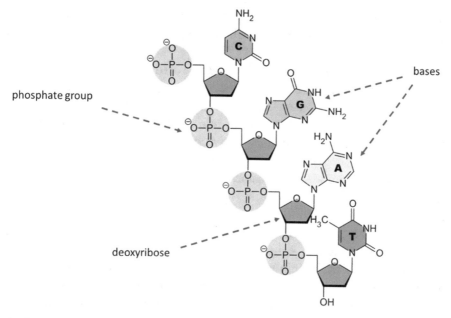

Figure 2.11 Structure of part of a deoxyribonucleic acid (DNA) molecule. The repeating unit is known as a nucleotide and the diagram shows four of these joined together. A single molecule of DNA will have millions of these nucleotides arranged in a specific sequence that is characteristic for a particular individual
Image modified from Sponk / Public domain

A protein (Figure 2.9) consists of amino acids (of which there are 21 different varieties in humans) linked together to form a long sequence which may consist of thousands of amino acids. When the number of amino acids is less than 50 the molecule is often called a "polypeptide" or "peptide" rather than a protein. A huge variety of proteins are found in humans and these differ because of the types of amino acids present and their sequence (which is dictated by our genes). Sugars may be attached to the protein chain to give a different type of macromolecule known as a glycoprotein. The long line of amino acids folds to form a particular shape which is characteristic of that particular protein and is very important in determining what function it can perform.

Polysaccharides consist of monosaccharides (sugars) joined together (Figure 2.10). They may form a linear or branched arrangement. Often, but not always, the polysaccharide contains only one type of sugar. Different polysaccharides arise because of differences in the type of sugar present as well as the arrangement of the sugars in the molecule. Sometimes the sugar molecule contains a phosphate, amino or sulphate group. Proteins can be attached to the polysaccharide to give rise to a different type of macromolecule known as a proteoglycan.

The deoxyribonucleic acid (DNA) molecule carries the genetic information of an organism and has a more complex structure (Figure 2.11) than that of a protein or polysaccharide. Although like these other two macromolecules, it's a polymer, the repeating unit is more complicated than an amino acid or a sugar. The repeating unit is known as a nucleotide and consists of three sub-units – a sugar (deoxyribose), a phosphate group and a nitrogenous base which may be adenine (A), cytosine (C), guanine (G) or thymine (T). The sequence of these bases is unique for every individual. The long chain of nucleotides arranges itself to form a helical structure.

Some macromolecules are even more complicated and consist of a protein as well as polysaccharide - these are known as proteoglycans and glycoproteins, both of which are important components of human tissues. A proteoglycan consists of a protein to which are attached chains of glycosaminoglycans (GAGs). A GAG is a polysaccharide that contains sugar molecules some of which contain an amino group (i.e. they are aminosugars) as well as uronic acid. They include chondroitin sulphate, dermatan sulphate, heparan sulphate and keratan sulphate. Glycoproteins also consist of a protein to which sugars are attached but in this case the proportion of sugar is much greater – approximately 50–60% compared to proteoglycans which usually contain less than 15% of sugar molecules.

Lipids are macromolecules with a very different type of structure to that found in the other macromolecules we've described because they aren't polymers. They consist of a glycerol molecule to which one or more fatty acids have been joined (Figure 2.12). Lipids are insoluble in water but dissolve in solvents such as ether and chloroform. They are classified as either fats (which are solid at room temperature) or oils (liquid at room temperature).

Now that we know a little about the structures of these important macromolecules, the next question to ask is where are they found in human tissues? The answer to this is that all of them are present inside the cells (Figure 2.13 and Box 2.2) that make up the tissues and many of them are also part of the extracellular matrix in which the cells of a tissue are embedded (Figures 2.5a and 2.15).

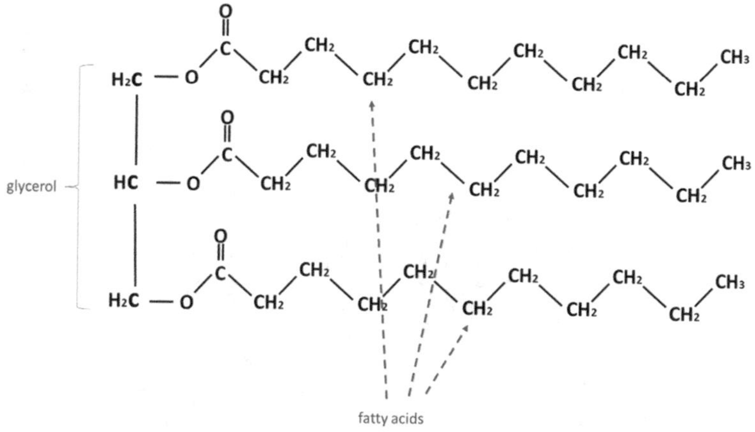

Figure 2.12 Diagrammatic representation of a lipid molecule. This consists of a glycerol molecule to which one or more (up to three) long-chain fatty acids have been attached

2.2.3 Where Are Macromolecules Found in Human Tissues?

Every human cell (Figure 2.13) contains a wide variety of compounds ranging from those with a low molecular mass to some with molecular masses in the millions.

Water comprises most of the mass of a human cell with proteins being the most abundant macromolecule (Figure 2.14).

The cytoplasm of the cell contains a wide range of low molecular mass compounds (such as simple sugars, amino acids, nucleic acids and fatty acids) as well as soluble and filamentous proteins. The nucleus contains mainly macromolecules including DNA, RNA and proteins. The structures within the cell are known as organelles (i.e. "little organs") and many of these (nucleus, lysosome, peroxisome, golgi body, mitochondrion etc. – Figure 2.13) are enclosed within a membrane which consists mainly of proteins and lipids. Ribosomes, the organelles responsible for making proteins, contain large proportions of RNA and proteins. Mitochondria are energy-generating organelles and contain a wide range of macromolecules including proteins (615 different types), lipids and DNA.

Each of the cells in a tissue contains about 10,000 different proteins and the total number of different proteins in the human body may be as high as 400,000. In addition to being present inside cells (intracellular proteins),

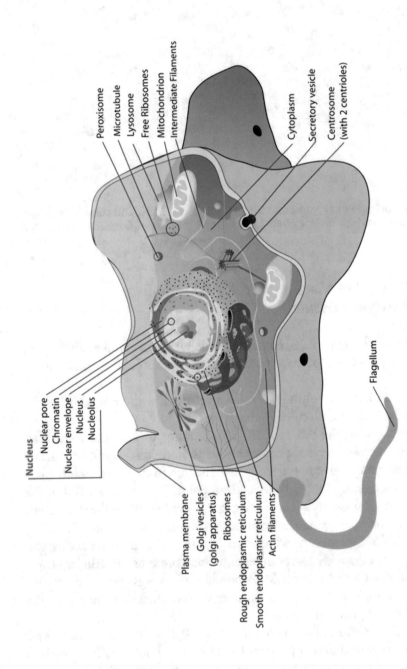

Figure 2.13 Diagram of a generalised human cell. This shows the main features of many of the cells that are present in a human being. However, the shape and size of human cells show considerable variation. For example, as pointed out previously, nerve cells are star-shaped and have thin extensions (axons) that can be up to one metre long. Red blood cells don't have a nucleus. Image courtesy of LadyofHats (Mariana Ruiz), public domain via Wikimedia Commons

Box 2.2 Human cells

The human body contains an incredible number of cells - about 4×10^{13}. This number is similar to the number of seconds in a million years. There are many different types of cells, each with a particular function and these include:

- Erythrocytes (red blood cells – Figure 2.5)
- Lymphocytes (white blood cells – Figure 2.5)
- Epithelial cells (Figure 2.2)
- Muscle cells (Figure 2.6)
- Nerve cells (Figure 2.7)
- Adipocytes (fat cells – Figure 2.5)
- Bone cells (osteoblasts and osteoclasts)
- Fibroblasts (Figure a.)

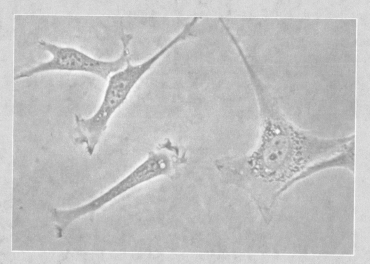

Figure (a) A group of four fibroblasts
SubtleGuest at English Wikipedia, CC BY 2.5 via Wikimedia Commons.

The most numerous cells are erythrocytes which account for 84% of the total number of cells. However, in terms of cell mass, muscle cells comprise the greatest proportion of the human body (Figure b). From the point of view of microbes and insects, it's the mass rather than the number that's the most significant in terms of the value of the human body as a source of nutrients.

Box 2.2 (continued)

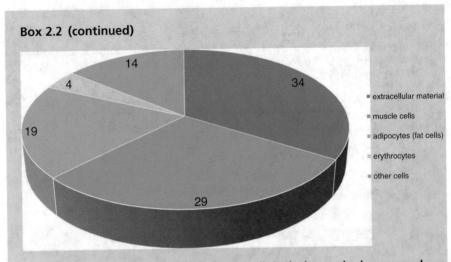

Figure (b) Proportions of the main types of cells in the human body expressed in terms of their mass. The figures show the % of the total body mass. Note that extracellular material (fluids, bone, extracellular matrix) accounts for about one third of the total body mass.

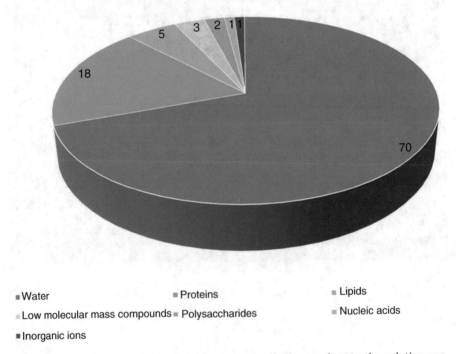

Figure 2.14 Chemical composition of a human cell. Figures denote the relative proportion (%) of each component

proteins (as well as glycoproteins and proteoglycans) are also found in the extracellular matrix (Figures 2.15 and 2.16) in which the cells are embedded – such macromolecules are termed "extracellular". The most important of these extracellular macromolecules are listed in Table 2.2. Collagen is the most abundant extracellular protein and comprises about one third of all the proteins in the human body – it's therefore a very important source of nutrients for microbes and insects.

As well as having a structural role, both inside and outside the cell, proteins are very important in metabolic processes because they function as enzymes i.e. they speed up the rate of chemical reactions. All of the chemical changes

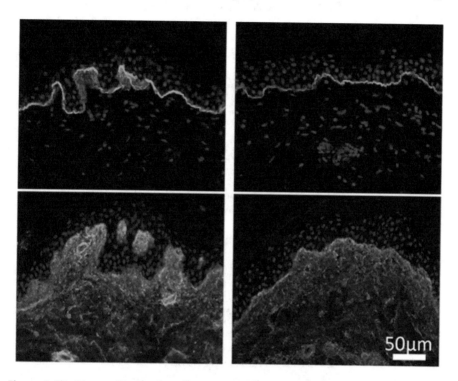

Figure 2.15 Photomicrographs of the connective tissue in facial skin from two sites (left and right images). The samples were stained to reveal the presence of laminin (upper images) and fibronectin (lower images). These two extracellular matrix proteins stain green. The nuclei of cells in the tissue stained red. Note that the fibronectin is more widely dispersed throughout the tissue than the laminin which appears as thin bands
Effects of a skin-massaging device on the ex-vivo expression of human dermis proteins and in-vivo facial wrinkles. Caberlotto E *et al. PLoS One.* 2017; 12(3): e0172624

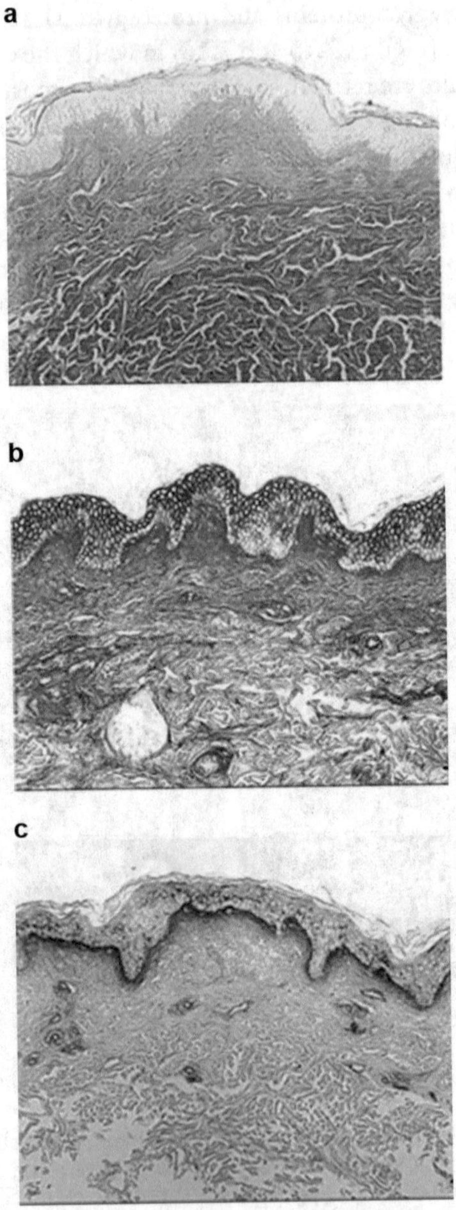

a

b

c

Figure 2.16 Cross-section through human skin which has been stained to show the presence of important constituents of the extracellular matrix. (a) collagen (stained red) and glycosaminoglycans (stained blue). (b) hyaluronic acid (stained brown). (c) chondroitin sulphate (stained brown)

Werth BB. *et al*. (2011) Ultraviolet Irradiation Induces the Accumulation of Chondroitin Sulfate, but Not Other Glycosaminoglycans, in Human Skin. *PLoS ONE* 6(8): e14830.

Table 2.2 Extracellular macromolecules found in human tissues

Proteins	Polysaccharides	Proteoglycans	Glycoproteins
Collagen, elastin, fibronectin, laminin, vitronectin	Hyaluronan, glycogen	These consist of a protein attached to a glycosaminoglycan (i.e. chondroitin sulfate, dermatan sulfate, heparan sulfate or keratan sulfate).	Mucins (found on the surfaces of cells and in the mucus layer that covers the gastrointestinal, respiratory and genitourinary tracts)

that occur within the body are only possible because of enzymes (Box 2.3). The decomposition of a human corpse is brought about mainly by enzymes. These come from the corpse itself and are also produced by microbes and insects (Chapters 3, 5 and 6).

Box 2.3 Enzymes – the workhorses of the cell

All of the chemical reactions involved in maintaining life in any organism are only possible because of enzymes. An enzyme is a catalyst i.e. it speeds up a chemical reaction so that it occurs in a fraction of the time that it would normally take. An enzyme is, therefore, a type of molecular machine – a nanomachine. It binds to a molecule and changes it in some way, or it binds to a number of molecules and joins them together. Enzymes, therefore, are involved both in the breakdown of molecules (known as catabolism – Figure a) and in the making of new molecules (known as anabolism – Figure b). They operate both inside and outside cells.

In this book we're concerned mainly with the ability of enzymes to break down macromolecules into smaller molecules that can then be used as nutrients by microbes and insects. This breakdown requires water and, therefore, is often termed "hydrolysis" because it involves water ("hydro") in the lysis (destruction) of a macromolecule. All enzymes have a name that ends in "ase" and this suffix is added to the type of molecule that the enzyme can modify in some way. For example a proteinase is an enzyme that acts on proteins, a lipase acts on lipids, a nuclease acts on nucleic acids and a polysaccharidase acts on polysaccharides.

Box 2.3 (continued)

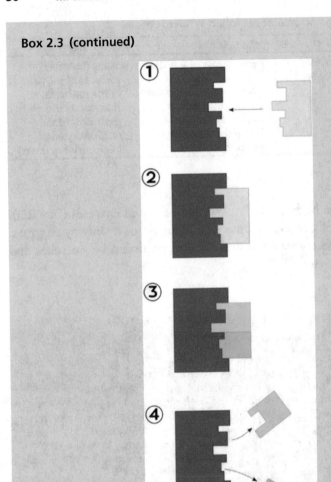

Figure (a) catabolic activity of an enzyme. A large molecule binds to a particular region (known as the "active site") of the enzyme molecule (2) and the enzyme converts it into several smaller molecules (3) which are then released (4) domdomegg, CC BY 4.0 <https://creativecommons.org/licenses/by/4.0>, via Wikimedia Commons.

Figure (b) Anabolic activity of an enzyme. Two molecules bind to the active site of the enzyme molecule (1, 2) and the enzyme joins them together to form a new molecule (3) which is the released (4)

Cells also contain several thousand different lipids and some are found extracellularly in plasma, muscles, arteries and skin. Inside the cell they are important constituents of membranes and are a source of energy. Outside the cell they function as important energy storage materials in the form of fats.

One of the most important polysaccharides present in human tissues is glycogen which is found inside cells as well as outside of them - mainly in the liver and muscles. It's a means of storing sugars when these aren't needed for immediate use. Hyaluronan (Figure 2.16) is an extracellular polysaccharide found mainly in connective tissue as part of the extracellular matrix which binds together, protects and supports the cells in a tissue. Proteoglycans are also major constituents of the extracellular matrix (Figure 2.16).

DNA is present in the nucleus and mitochondria of a cell while ribonucleic acid (RNA) is present in both the nucleus and the cytoplasm.

2.3   How Available to Microbes and Insects Are the Nutrients Present in the Human Body?

From the above description, we can appreciate that the human body is a rich source of a wide range of potential nutrients, but how accessible are they to the two main types of organism responsible for its decomposition – microbes and insects? Low molecular mass constituents of human tissues don't need any processing by microbes or insects and so are immediately available to both types of organism. They are simply absorbed into the creature where they provide energy or can be used as building blocks to make new macromolecules. A wide range of such compounds (more than 110,000) are present in the blood (Box 2.4), tissue fluid (the extracellular liquid that bathes the cells of most tissues), sweat, saliva and in the contents of the gastrointestinal tract. Such compounds include glucose, glycerol, lactic acid, acetic acid, citric acid, phosphoric acid, cholesterol, urea, amino acids and vitamins. All of these will be immediately available to other organisms as soon as you die. Similarly, ions such as Sodium, Potassium, Magnesium, Calcium, Chloride, Sulphate etc., can be absorbed directly into the cells of the creatures that feed on our corpse.

Box 2.4 Blood and tissue fluid as nutrient sources

These two types of fluid are present in large quantities in an adult, are highly nutritious and are completely free of microbes. They therefore constitute an important source of nutrients for microbes and insects.

The average adult has between 4 and 5 litres of blood. About 45% of blood consists of cells and the remaining fluid is known as plasma. We've already seen that cells are a rich source of nutrients, but what about plasma? Plasma is made up mainly of water (about 91%) and proteins (about 8%), the rest consists of minerals and a variety of small molecules. More than 4000 different compounds have been detected in plasma. The main proteins found are fibrinogen (involved in blood clotting), albumin and globulins (antibodies and enzymes). The minerals present include sodium, potassium, calcium, chloride and bicarbonate. The small molecules include important nutrients such as amino acids, carbohydrates, urea and lipids. Plasma is therefore a valuable source of nutrients because not only is there a lot of it but also it has a high content, and variety, of proteins and useful small molecules.

All of the cells in our tissues are surrounded by an aqueous liquid known as tissue or interstitial fluid. This fluid is produced mainly as a result of leakage from the blood capillaries. The total volume of tissue fluid in an adult is about 11 litres. It has a similar composition to plasma except that its protein content is lower because many plasma proteins can't pass through the walls of the blood capillaries. However, other proteins, shed by cells, are usually present. As many as 670 different proteins and almost 200 different lipids have been found in tissue fluid, although their concentrations are generally lower than those present in plasma. The smaller molecules present include amino acids, carbohydrates, urea and various minerals.

2.4   And then, Of Course, We Shouldn't Forget That Brown, Smelly Stuff

Adults carry approximately 250 g of faecal material in their large intestines and, once we have died, this will become available to microbes, insects and other animals. About 75% of faeces is water, the rest consists mainly of microbes, proteins, carbohydrates, lipids and minerals (Figure 2.17).

We'll be describing the microbes present in faeces in Chapter 3. The proteins consist of those in the diet that haven't been digested, ones from dead microbes and those present in dead intestinal cells. The main carbohydrates are those from plants that were part of the diet and haven't been digested – this is commonly referred to as "fibre". The fats present come mainly from the diet, dead microbes and dead intestinal cells. Calcium, iron and phosphate are the dominant minerals present. Human faeces, therefore, has a high

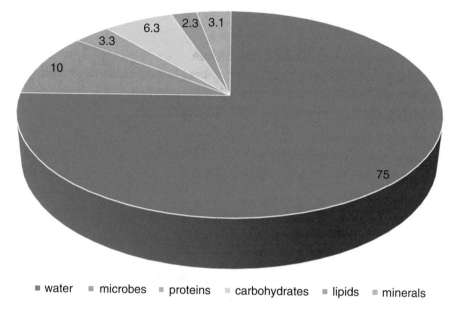

■ water   ■ microbes   ■ proteins   ■ carbohydrates   ■ lipids   ■ minerals

Figure 2.17 Proportions (%) of the main constituents of adult human faeces. The figures shown should be regarded as only approximations because the exact composition varies tremendously depending on the nature of the individual's diet, physiology and physical activities

nutritional value – mainly in the form of macromolecules. As we'll see in Chapter 6, many insects are highly appreciative of this and regard faeces as the ideal place to lay their eggs so that the newly-hatched larvae have an adequate supply of nutrients. Many of the volatile compounds (e.g. ammonia, carboxylic acids, indole, sulphur-containing compounds) produced by the microbes present in faeces can attract insects.

To summarise, a human corpse is a very rich source of a wide variety of nutrients. Many of these are present as small molecules that can be used directly by a wide range of creatures, both microscopic and macroscopic. However, most of the nutritional potential of a human corpse is tied up in macromolecules and these must first of all be broken down to their small molecular mass components in order to be used by the microbes and insects that bring about the recycling of our tissues. This is achieved by the activity of an extensive range of enzymes. These enzymes are produced, remarkably, by our own cells as well as by microbes and insects. The way in which macromolecules are broken down will be described in more detail in Chapter 5.

2.5   Want to Know More?

Blausen Scientific and Medical Animations
An amazing collection of 645 videos describing human anatomy and physiology from the chemical to the tissue level of organisation.
https://blausen.com/en/topic/anatomy--physiology/

Innerbody research
Detailed descriptions and diagrams of the various structures of the human body.
https://www.innerbody.com/htm/body.html

Get body smart
An online examination of human anatomy and physiology
https://www.getbodysmart.com/

Human tissues
https://open.oregonstate.education/aandp/chapter/4-1-types-of-tissues/
https://www.studyread.com/types-of-tissues/

Human physiology
https://microbenotes.com/category/human-physiology/

Introduction to macromolecules
Descriptions, diagrams and videos of the main types of macromolecules - from the Khan Academy
https://www.khanacademy.org/science/biology/macromolecules

Biological macromolecules
Descriptions and diagrams of the main types of macromolecules provided by OpenStax
https://bio.libretexts.org/Bookshelves/Introductory_and_General_Biology/
Book%3A_General_Biology_(OpenStax)/1%3A_The_Chemistry_of_
Life/3%3A_Biological_Macromolecules

Enzymes
A section from the "Understanding Chemistry" online resource provided by Chemguide. A teaching site for students at the pre-university level
https://www.chemguide.co.uk/organicprops/aminoacids/enzymes.html

Enzymes and digestion
Descriptions, diagrams and a video about digestive enzymes provided by the BBC.
https://www.bbc.co.uk/bitesize/guides/zwnstv4/revision/1

Enzymes: principles and biotechnological applications
A more advanced description of enzymes and how they function
Essays in Biochemistry 2015 Nov 15; 59: 1–41. Peter K. Robinson
https://www.ncbi.nlm.nih.gov/pmc/articles/PMC4692135/

The Human cell
Section of a course on human physiology by Gary Ritchison, Foundation Professor, Department of Biological Sciences, Eastern Kentucky University
http://people.eku.edu/ritchisong/301_notes_1.htm

Introduction to the cell – secondary school student level
http://biologymad.com/resources/Ch%201%20-%20Cells.pdf

A series of 3-D animations of the human cell from the Smithsonian National Museum of Natural History and the National Human Genome Research Institute.
https://unlockinglifescode.org/3d-animations-of-the-human-cell

3

Our Life-Long Microbial Companions – Who Are They and What Do They Get Up to While We're Alive?

For the first few hours after death, before any insects arrive, the main drivers of decomposition of the human body are autolysis and the microbes that live on us (known as the "human microbiota" or "human microbiome"). It shouldn't really come as any surprise to learn that your body is a home to a whole bunch of microbes. After all, we live on a planet that's densely populated with these creatures – estimates suggest there are approximately 10^{31} of them. That's an incredible number. To give you an idea of just how big that number is, it's been estimated that the total number of stars in the universe is 10^{24} – so there are 10 million times more microbes on our planet than there are stars in the universe. Astonishingly, 18% of the total mass of living creatures (the "biomass") on our planet consists of microbes whereas animals amount to less than 1%. Not only are there huge numbers of microbes but there's also an enormous variety – at least 10^{12} different species. Microbes are everywhere - in the air we breathe, on the objects we touch and in the food we eat. They've colonised virtually all parts of our planet, from the depths of the ocean to the tops of mountains, so it's no surprise that some of them have made their homes on our bodies.

Autolysis (the self-destruction of a cell) has been mentioned in Chapter 1 and will be described in more detail in Chapter 5. The purpose of this chapter is to introduce you to the human microbiota. Now that you've absorbed the information in Chapter 2 and have become a biochemist, we're now going to turn you into a microbiologist as well – you're going to be so interesting at dinner parties!

© The Author(s), under exclusive license to Springer Nature Switzerland AG 2022
M. Wilson, *Life After Death: What Happens to Your Body After You Die?*,
Springer Praxis Books, https://doi.org/10.1007/978-3-030-83036-6_3

3.1   What Are Microbes?

In order to understand the central role that microbes play in the decomposition of a human corpse, it's essential to know which microbes are present on the body of a human at the time of death and what they're capable of doing to a corpse. But, first things first. It's important at this point to get clear just what we mean by the term "microbe". There's a lot of confusion in the popular press and other media sources about this word, and reporters and commentators tend to use the terms microbe, bacterium, virus and bug interchangeably and unthinkingly. Does this matter? Yes, it certainly does – there are enormous differences in the way in which bacteria and viruses, for example, behave and what they're capable of doing to a human body – whether it's alive or dead.

3.1.1   What Is the Definition of a Microbe?

The original definition of a microbe was an organism that could only be seen with the help of a microscope. There are six main types of microbe – bacteria, viruses, fungi, protozoa, archaea and algae. While it's generally true that we do need a microscope to see the vast majority of microbes, there are exceptions. For example the mushrooms we see in the fields and those that we eat are fungi but they're certainly not microscopic (Figure 3.1a). Also, the seaweeds that we find on our visits to the seaside are, in fact, various types of algae (Figure 3.1b). These are examples of microbes that form such large multicellular structures that they become visible to the naked eye.

3.1.2   So, How Are These Six Types of Microbes Different From Each Other?

Of the six types of microbes listed, only the first five are found on the human body – algae are photosynthetic microbes that live mainly in aquatic environments and don't colonise humans. The five types of microbe that do live on us are very different and interact with the human body in distinctive ways. Their main characteristics are summarised in Table 3.1 and typical examples of each type are shown in Figure 3.2. As you can see from Table 3.1, the main types of microbe differ considerably from one another. What's guaranteed to annoy a microbiologist and a fledgling microbiologist like yourself, is the constant misuse of these terms. For example anyone who said that an oak was a type of fish would be laughed at. Yet, reporters and commentators often refer to a bacterium such as *Staphylococcus aureus* as being a type of virus – ridiculous isn't it?.

(a)

(b)

Figure 3.1 Examples of microbes that we can see without the help of a microscope.
(a) a mushroom – a type of fungus (b). a seaweed – a type of alga
(a) Actia Nicopolis Foundation, CC BY 4.0
(b) <https://creativecommons.org/licenses/by/4.0>, via Wikimedia Commons

3.1.3   They're So Small, So They Must Be Very Simple Creatures?

You shouldn't be deceived by appearances. Although many microbes, particularly bacteria and archaea, look as though they're very simple, higher magnification shows them to have very complex structures. Their complexity has been revealed by very powerful microscopes such as the electron microscope which can magnify things many thousands of times (Figure 3.3).

Table 3.1 The main characteristics of the five types of microbe found on humans. Although most microbes are too small to be seen with the naked eye, some form structures that are visible to us e.g. mushrooms, seaweeds. The size of a microbe is usually expressed in micrometres. One micrometer (μm) is a millionth of a metre

Type of microbe	Size	Main features
Bacterium	About 1-2 micrometres in diameter	• Exist as a single cell (i.e. Unicellular) • Reproduce by splitting in two • Most have a cell wall • Some are motile • A few can form spores • Genetic material is DNA
Virus	Usually less than 0.5 micrometres in diameter	• Don't have a cellular structure • Can't reproduce by themselves • Genetic material can be DNA or RNA • Aren't motile
Fungus	Most are much larger than bacteria. They are several micrometres in diameter and can be very long (some are very large e.g. mushrooms)	• Usually multicellular and form filaments • Have a cell wall • Reproduce by forming spores • Genetic material is DNA
Protozoan	Most are much larger than bacteria; often tens of micrometres in diameter	• Unicellular • Reproduce by splitting in two • Don't have a cell wall • Have a flexible shape • Most are motile • Genetic material is DNA
Archaeon	About 1-2 micrometres in diameter	• Unicellular • Have a cell wall • Reproduce by splitting in two • Genetic material is DNA • Most are found in extreme environments such as hot springs etc

3.1.4 How Do We Identify the Various Types of Microbes?

It's easy for us to tell the difference between an elephant and a daffodil and to identify a particular type of flower or animal. This is all done on the basis of their appearance and this is the usual way we identify large organisms. Because most microbes are so small, we have to use a microscope to actually see them. Usually,

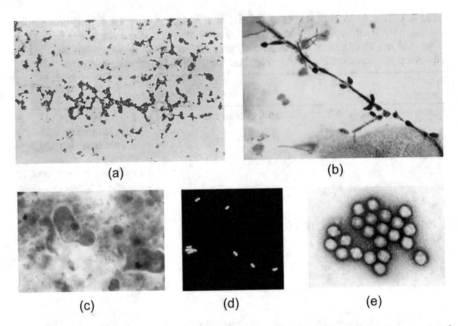

Figure 3.2 Microscopic appearance of typical examples of the five main types of microbe that are found living on healthy humans. Note that the vast majority of microbes are not coloured and these images show microbes that have been stained by dyes (to make them easier to see) or else the photos have been digitally colourised

(a) Photomicrograph of *Staphylococcus aureus*, a bacterium found in the nostrils (magnification X1250). It consists of round cells, approximately 1.0 μm in diameter. As can be seen in this image, they generally occur in groups consisting of many cells rather than as individual cells

Image courtesy of Dr. Richard Facklam, Centers for Disease Control and Prevention, USA

(b) Photomicrograph of *Candida albicans*, a fungus that's often present in the mouth and colon of healthy individuals (X1000). The organism exists as oval cells that also produces long filaments

Image courtesy of Dr. Stuart Brown, Centers for Disease Control and Prevention, USA

(c). Photomicrograph of *Entamoeba coli*, a protozoan often found in the colon of healthy individuals (X1150). The usual size of the organism is 15-50 μm and so is much larger than a bacterium

Image courtesy of Dr. Green, Centers for Disease Control and Prevention, USA

(d) Photomicrograph of a *Methanobrevibacter* species which is an archaeon. Archaea are similar in size to bacteria and have a diameter of approximately 1.0 um.

The complete genome sequence of the rumen methanogen *Methanobrevibacter millerae* SM9. Kelly WJ *et al.*

Standards in Genomic Sciences 2016; 11: 49. This article is distributed under the terms of the Creative Commons Attribution 4.0 International License (http://creativecommons.org/licenses/by/4.0/)

(e) Most viruses are too small to be seen through an ordinary microscope and we have to use an electron microscope. This image is an electron micrograph of an adenovirus. 22 virus particles (known as "virions") are shown, each of these has a diameter of approximately 100 nm which is about one tenth of the diameter of a typical bacterium. Adenoviruses are often present in the respiratory tract

Image courtesy of Dr. G. William Gary, Jr. Centers for Disease Control and Prevention, USA

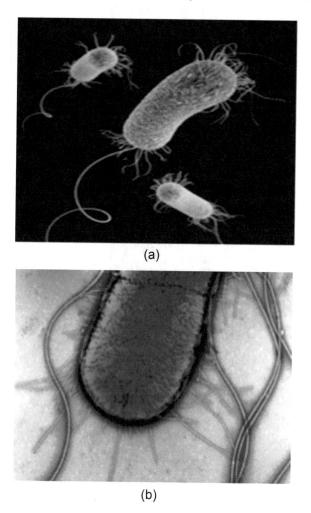

(a)

(b)

Figure 3.3 Electron microscopy shows the complexity of even the simplest microbes
(a) This is a three-dimensional computer-generated image of a bacterium. The artistic recreation was based upon images taken with an electron microscope. Note the presence of many narrow, thread-like structures – these are fimbriae that help the microbe to stick to cells and surfaces. Also a long, corkscrew-shaped structure (a flagellum), which enables the bacterium to move
Jennifer Oosthuizen - Medical Illustrator, Centers for Disease Control and Prevention, USA
(b) Part of a bacterial cell as seen through an electron microscope (magnification approximately 50,000 times). This shows the dark-staining cell wall, details of the internal structure, the long, whip-like flagella and shorter, hair-like fimbriae
David Gregory & Debbie Marshall. CC BY 4.0

looking at them through a microscope gives us a good idea as to which of the six major groups a particular microbe belongs. In the case of most protozoa, fungi and algae, their appearance through the microscope is also usually enough to distinguish between the various species within these groups. So, if we have found a particular protozoan, fungus or alga in a sample then looking at it through a microscope is usually enough to identify which species it is. Unfortunately, this isn't the case with bacteria, archaea or viruses. It's been estimated that there are as many as 10^9 different species of bacteria on planet Earth but all of these have one of three basic shapes – spherical, rod or spiral (Figure 3.4). So, we certainly can't identify a particular bacterium on the basis of its appearance alone.

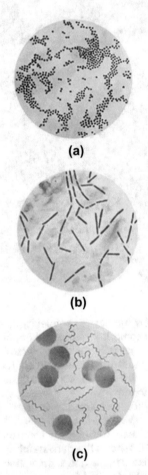

(a)

(b)

(c)

Figure 3.4 Drawings of photomicrographs showing the three main shapes of bacteria: (a) this shows a very large number of dark blue spheres in clumps – these are known as cocci, (b) about 40 long, thin, blue rods can be seen – these are known as bacilli and (c) about 15 red, snake-like structures are present – these are spirillar (spiral-shaped) bacteria. The 7 much larger, round structures are red blood cells
All three images courtesy of Centers for Disease Control and Prevention, Atlanta, USA

The identification of archaea and viruses presents a similar problem – a huge number of species in each case, but only a very limited range of shapes. How we identify bacteria, as well as other microbes, is outlined briefly in Box 3.1.

Box 3.1 Identification of bacteria and other microbes

Once we've looked at a bacterium through a microscope and know its shape, the next most important test used to identify it is what it looks like after it's been stained in a procedure developed by a Danish scientist, Christian Gram in 1884. This involves putting it onto a glass slide and treating it with a sequence of purple and red dyes. Those bacteria that end up looking purple are called "Gram-positive" while those that are red are "Gram-negative" (Figure a).

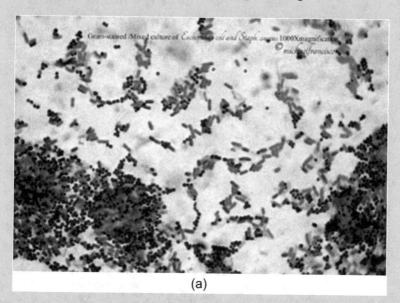

(a)

Figure (a) Photomicrograph of a Gram stain of a mixture of Gram-positive cocci (*Staphylococcus aureus* - purple) and Gram-negative bacilli (*Escherichia coli* - red). Magnification x1,000
Michael R Francisco from France / CC BY (https://creativecommons.org/licenses/by/2.0)

Although this staining procedure was invented more than a century ago, it's still an important step in bacterial identification. Other important characteristics used in identifying a bacterium are how it grows, what it feeds on and what waste materials it produces. Important questions include:

* does it need oxygen to grow?
* can it move?

Box 3.1 (continued)

- does it form spores?
- what compounds can it use as nutrients?
- can it break down macromolecules?
- at what temperatures can it grow?
- can it grow in acidic conditions?
- can it grow in high concentrations of salt (sodium chloride),
- what waste products does it produce?

These characteristics, as well as many others that haven't been listed, are also important in determining where on planet Earth the bacterium can live i.e. its habitat. Just as a rose needs a certain type of soil and particular climatic conditions (temperature, rainfall etc.) in order to grow, so does each type of microbe. Growing a bacterium in the laboratory is an important stage in identifying it and this is done by transferring it to a petri dish (Figure b) containing a jelly-like material that contains all of the nutrients it needs – this is known as a "medium". The petri dish is then put into an incubator (and usually left overnight) which keeps it at a temperature (usually 37 °C for bacteria isolated from humans), and in an atmosphere suitable for its growth. Each bacterium grows and reproduces until it produces a visible structure known as a colony.

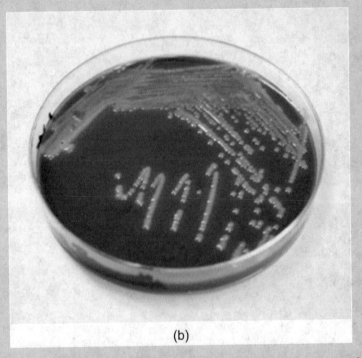

(b)

Figure (b) Plastic petri dish containing a nutrient medium. In this image, the lid of the petri dish has been removed and colonies of a bacterium growing on the surface of the medium can be clearly seen
Megan Mathias and J. Todd Parker, Centers for Disease Control and Prevention, USA

Box 3.1 (continued)

Although the above procedures are still widely used to identify bacteria they're being replaced by methods based on analyzing their genetic material (DNA) – these are referred to as "molecular techniques" or "sequencing" (because they determine the sequence of the DNA of the bacterium). Such molecular techniques are also increasingly being used to identify archaea, fungi and viruses. Fungi and protozoa are still identified mainly on the basis of their microscopic appearance. Many viruses can be identified on the basis of their appearance when viewed through an electron microscope.

3.1.5 Meet My Best Friends - My Microbiota

Until the moment of our death, we'll have lived throughout our lives with an enormous number and variety of microbes which, collectively, are known as the "human microbiota", "human microbiome" or "indigenous microbiota". To parody that great rock-and-roller Cliff Richard, we are a "crying, talking, sleeping, walking, living…… incubator of microbes".

While we're alive, this is generally a harmonious relationship, and is described as being a symbiosis (Figure 3.5) which means that both our microbiota and ourselves benefit from the association. We provide our microbiota with a safe environment and supply warmth, moisture and nutrients. They, in turn, protect us from dangerous pathogens, help our immune system to develop, help us to digest our food and also provide us with energy and vitamins (see Box 3.5). There are microbial communities on all of those surfaces of our body that are in contact with the external environment. These obviously include the skin and mouth but, less obviously, the respiratory tract, the gastrointestinal tract and the genitourinary tract (Figure 3.6). Microbes aren't usually present in our internal tissues and organs such as the heart, liver, brain etc. – this only happens when we suffer from certain types of infection.

3.1.6 How Many Microbes Live on Me?

The number of microbes present on a human being is staggering and it's been estimated that there are at least 10^{14} bacteria (which is 100,000,000,000,000), 10^{14} viruses and an unknown number of fungi, archaea and protozoa on a single adult. Furthermore, the microbial communities found on various parts of our body are extremely complex and altogether more than 20,000 different microbial species have been detected.

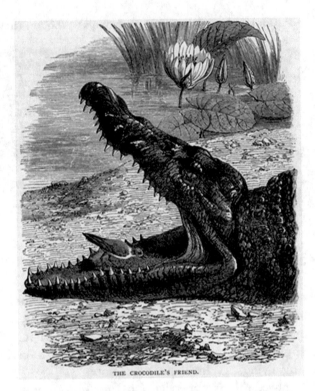

THE CROCODILE'S FRIEND.

Figure 3.5 An example of a symbiotic association between a Nile crocodile and a Plover. The plover eats leeches and other insects trapped in the crocodile's teeth, while the crocodile benefits by having its teeth cleansed of such parasites
Henry Scherren, Public domain, via Wikimedia Commons

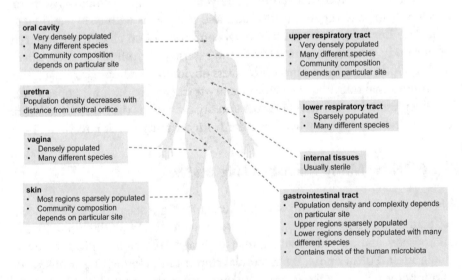

oral cavity
- Very densely populated
- Many different species
- Community composition depends on particular site

urethra
Population density decreases with distance from urethral orifice

vagina
- Densely populated
- Many different species

skin
- Most regions sparsely populated
- Community composition depends on particular site

upper respiratory tract
- Very densely populated
- Many different species
- Community composition depends on particular site

lower respiratory tract
- Sparsely populated
- Many different species

internal tissues
Usually sterile

gastrointestinal tract
- Population density and complexity depends on particular site
- Upper regions sparsely populated
- Lower regions densely populated with many different species
- Contains most of the human microbiota

Figure 3.6 The main sites in a human that are colonized by microbes. The microbial communities found at these sites are collectively known as the "human microbiota" or "human microbiome"

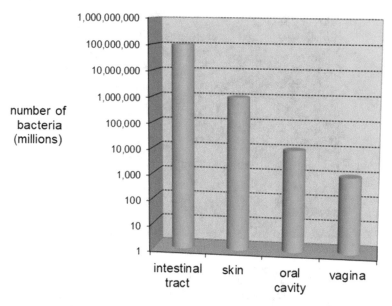

Figure 3.7 An estimate of the numbers of bacteria present at the most heavily popu-lated sites in the human body. Remember that these figures show only the numbers of bacteria. In addition there are probably similar numbers of viruses at each site. We have little idea of the numbers of fungi, archaea and protozoa at each body site

Although, as mentioned, all of the body's surfaces support microbial com-munities, most of the human microbiota (in terms of numbers - Figure 3.7) is found in the colon, the lower region of the gastrointestinal tract (GIT). However, the skin and mouth also have substantial numbers of microbes. While the respiratory and genitourinary tracts are also heavily colonized by microbes, the total numbers present are small compared to those in the GIT and on the skin and so there's no need to discuss these in detail.

3.1.7 Do All Parts of My Body Have the Same Microbes?

It's important to realise that each body site has its own distinctive microbial community. This means that we can talk about the "oral microbiota", the "skin microbiota" etc. These are all very different from one another. The rea-son why they differ is because the environment at each body site (known as a "habitat" – Figure 3.8) provides a unique set of conditions which allows only certain microbes to grow there. In other words, we can explain these differ-ences in terms of ecology – but this is ecology on a microscopic scale (micro-ecology) rather than on the large scale (macroecology) that we are more

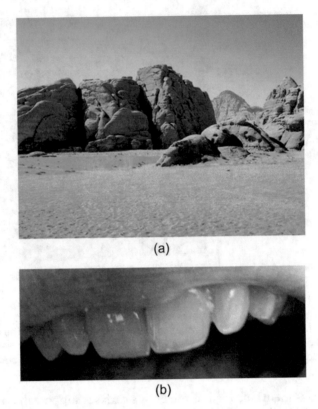

(a)

(b)

Figure 3.8 The nature of the environment affects what creatures can live there. (a) A desert; the normal habitat of a camel, but we wouldn't expect to find a dolphin living there. (b) Human teeth; the normal habitat of the bacterium *Streptococcus mutans* but we won't find any *Escherichia coli* living there
(a) Peter Chisholm, CC0, via Wikimedia Commons

familiar with in our daily lives. We can easily understand why a giraffe wouldn't be found living in the frozen wastes of the Antarctic and we'd be surprised to see a penguin waddling along in the Sahara desert. For similar reasons (i.e. ecological factors, but on a microscopic scale), we don't find bacteria such as *Bacteroides* species on the skin or *Clostridium* species in the lungs.

So, let's take a look at those body sites that have large microbial communities. We'll describe the main ecological features of each of these habitats and the composition of each of the communities that live there - for a poetic summary of the nature of these communities see Box 3.2. Importantly, we'll also learn about how these communities are able to play a role in the decomposition of a human corpse.

Box 3.2 Microbiota Row

For fans of Bob Dylan (can there be anyone who isn't?), this is a parody of one of his best songs "Desolation Row". It describes the main types of microbes found on the various parts of your body and should be sung to the rhythm of the original song.

The eyes have a microbiota
That's very sparse indeed
A few Gram-positive cocci
Scavenge from tears all that they need
The skin has a denser population
P. acnes is plentifully found
While coryneforms and staphylococci
Are invariably around.
But molecular tools have shown us
There's much more still to know
About the microbes that live upon us
And even help us grow.
The respiratory tract is moist and inviting.
With food aplenty there
But of the mucociliary escalator
All microbes must beware
Yet haemophili and streptococci
And Neisseria can survive.
While Mollicutes and Moraxella.
Will there be found alive.
But there are pathogens among them.
Most deadly, that's for sure
Armed with many deadly toxins
To bring us to death's door.
Inside the terminal urethra
Staphylococci hold on tight
But most of the urinary tract is sterile
Thanks to innate immunity's might.
A male's reproductive system is arid
But with microbes a female's abounds
With lactobacilli and other genera
Their variety astounds.
But hormones have a great effect
On which microbes there can grow
And their relative proportions
Change as time's stream does flow.
From the mouth down to the rectum.
The intestinal tract unwinds
Producing ecosystems so complex
And microbiotas of many kinds.
The oral cavity is aswarming
With 800 taxa there
While the hostile, acidic stomach

Box 3.2 (continued)

Apart from *H. pylori* is almost bare.
The small intestine is nearly sterile
But the colon is replete
With almost a thousand species
And without them we're incomplete.
Yes, I know you think they're nasty
Those minutest forms of life
Your mother said that they were dirty
And would only cause you strife.
But they're essential for your survival.
Believe me you really must
They digest our food and protect us from
Pathogens that would make us dust.
Most of our indigenous microbes
Play a beneficial role, so please
Don't disturb or try to remove them
The result would be disease.

Michael Wilson, 2007

3.1.8   Which Microbes Live in My Gut?

The digestive system consists of the gastrointestinal tract (GIT) together with what are known as the accessory digestive organs such as the teeth, tongue, salivary glands, liver, gallbladder and pancreas. Its main function is to convert the constituents of our diet into small molecules and then to absorb these for subsequent distribution throughout the body by the bloodstream. The GIT consists of several distinct regions – the oral cavity, the pharynx, oesophagus, stomach, small intestine (duodenum, jejunum, and ileum) and the large intestine (caecum, colon, and rectum). Essentially, it can be considered to be a continuous tube (although it's very complex and convoluted) that extends from the mouth to the anus (Figure 3.9).

The environments within the various regions of the GIT are very different which means that each has its own distinctive microbiota.

3.1.8.1   Microbes That Live in Our Mouth

The mouth is anatomically quite complicated and contains a large variety of structures (Figure 3.10), all of which are covered in microbes. However, the vast majority of oral microbes live in large aggregates (known as biofilms) on

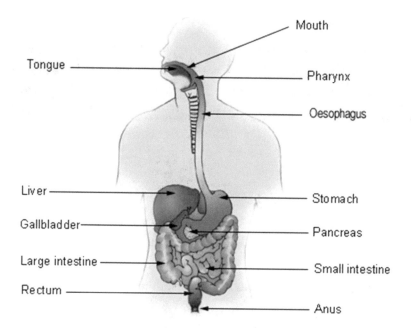

Figure 3.9 Diagram showing the main organs of the digestive system, which consists of the gastrointestinal tract and the accessory digestive organs
From http://www.training.seer.cancer.gov/module_anatomy/anatomy_physiology_home. html; funded by the US National Cancer Institute's Surveillance, Epidemiology and End Results (SEER) Program with Emory University, Atlanta, Georgia, USA

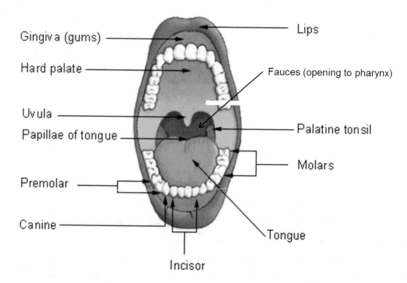

Figure 3.10 Diagram showing the main features of the oral cavity
From http://www.training.seer.cancer.gov/module_anatomy/anatomy_physiology_ home.html; funded by the US National Cancer Institute's Surveillance, Epidemiology and End Results (SEER) Program with Emory University, Atlanta, Georgia, USA

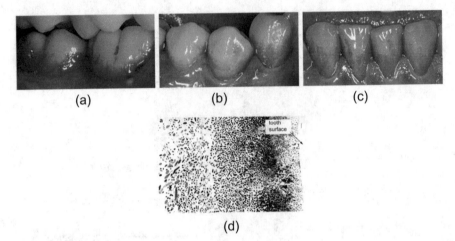

(a)   (b)   (c)

(d)

Figure 3.11 Images of dental plaque – a biofilm that accumulates on the surface of teeth

(a–c). Dental plaque can be revealed by using a coloured "disclosing" agent. Plaque is white or yellowish and so is difficult to see on the surface of white teeth. However, it's possible to show its presence by using a mouthwash containing a dye that sticks to the plaque. These are examples of the use of a red disclosing agent to reveal the distribution of plaque in three individuals

Clinical validation of robot simulation of toothbrushing--comparative plaque removal efficacy. Lang T. *et al.,- BMC Oral Health*. 2014 Jul 4; 14:82. https://doi.org/10.1186/1472-6831-14-82 This is an Open Access article distributed under the terms of the CreativeCommons Attribution License (https://doi.org/creativecommons.org/licenses/by/2.0), which permits unrestricted use, distribution, and reproduction in any medium, provided the original work is properly credited

(d). Electron micrograph of dental plaque from a healthy volunteer. Bar in upper right hand corner = 3 μm

Image kindly supplied by Mrs. Nicola Mordan, UCL Eastman Dental Institute, University College London

the surfaces of the teeth (Figure 3.11). We usually call these biofilms dental plaque.

The tongue is also heavily colonised with microbes because it is covered in a large number of small, lumpy structures (known as papillae) and therefore forms a very convoluted surface with lots of microscopic valleys within which microbes can shelter and grow (Figure 3.12).

The oral microbiota has a plentiful supply of nutrients from saliva which contains proteins, glycoproteins, carbohydrates, amino acids and lipids. We produce approximately 750 ml of saliva per day. Oral microbes can also benefit from the wide range of nutrients that are present in the food that we chew several times each day. Because it's open to the atmosphere, the mouth has

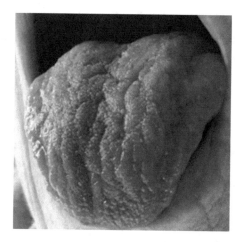

Figure 3.12 Human tongue. The tongue is covered in a large number of small protrusions (papillae) as well as smaller numbers of taste buds. It has, therefore, a very convoluted surface

plenty of oxygen to enable the growth of those microbes which, like us, need oxygen to survive (see Box 3.3). However, there are also regions in which the oxygen supply is more limited such as the gaps between the teeth and between the papillae of the tongue, as well as in the depths of dental plaque. These

---

Box 3.3 Microbes and oxygen

Many microbes, like humans and all of the other animals on this planet, need oxygen to survive – such organisms are known as "aerobes" (from the Greek words "aer" and "bios" which mean air and life respectively). Oxygen is used mainly to produce energy (in the form of the molecule adenosine tri-phosphate – ATP) by a process known as respiration. Remarkably, however, many microbes can also produce ATP without the need of oxygen – this is known as "anaerobic respiration" (the word anaerobe comes from "an" meaning "without" and "aerobe"). ATP can also be produced in many organisms by a process known as fermentation which involves the conversion of sugars and other organic compounds to simpler molecules such as lactic acid and alcohol.

On the basis of their need for oxygen, it's possible to classify microbes into three main groups:

(i) Aerobes – these need oxygen to grow and reproduce
(ii) Anaerobes- these don't need oxygen to grow and reproduce
(iii) Facultative anaerobes – these can grow and reproduce in the presence or absence of oxygen

This ability to grow in the presence or absence of oxygen makes microbes, as a group, extremely versatile. We'll see later that anaerobic growth of microbes is very important in the decomposition of the human body.

regions provide conditions suitable for the growth of anaerobic microbes and, surprisingly, the oral microbiota contains very high proportions of such microbes. As we shall see in Chapter 5, anaerobic microbes are very important in the decomposition of the human body.

When it comes to describing a microbial community at a body site, it's important to consider two features of that community – how often a particular microbe is found there in different people and what proportion of the community it comprises when it's present there. For example, a particular microbe may be regularly found at a body site but could comprise only a small proportion of that community when it's present. Dental plaque contains huge numbers (about 10^8 per mg) and a large variety (at least 800 species) of bacteria. It almost always contains bacteria belonging to the genera *Streptococcus, Corynebacterium, Actinomyces, Fusobacterium, Veillonella* and *Prevotella* (Figure 3.13).

Streptococcus species are usually present in the highest proportion in dental plaque (Figure 3.14). Streptococci and corynebacteria are facultative anaerobes whereas *Actinomyces, Fusobacterium, Veillonella,* and *Prevotella* are all anaerobes. Fungi, mainly *Candida* species (Figure 3.2b), and viruses (approximately 10^7 per mg of plaque) are also usually present. The naming system used for bacteria and other microbes can be confusing and so it's worth taking a look at Box 3.4 below which explains this in a little more detail.

Collectively, oral microbes produce a wide range of proteases, peptidases, sialidases (enzymes involved in breaking down mucins and other glycoproteins) and glycosidases (also known as polysaccharidases) that can break down polysaccharides. Consequently, they're able to break down a wide variety of the proteins, glycoproteins and polysaccharides that are present in the human body (Table 3.2).

Those genera that contain species able to break down polysaccharides include *Streptococcus, Fusobacterium, Propionibacterium* and *Corynebacterium*.

Many oral microbes can degrade DNA including members of the genera *Streptococcus, Fusobacterium, Lactobacillus, Porphyromonas, Actinomyces, Leptotrichia, Prevotella, Capnocytophaga, Aggregatibacter, Tannerella* and *Campylobacter*.

Lipases are produced by species of the genera *Streptococcus, Capnocytophaga, Campylobacter* and *Prevotella*.

Fungi are also present in the mouth and the most frequently-detected types are shown in Figure 3.15.

Fungi found in at least 50% of individuals are *Candida* (Figure 3.2b), *Cladosporium, Aureobasidium* and members of the order *Saccharomycetales* (Figure 3.16).

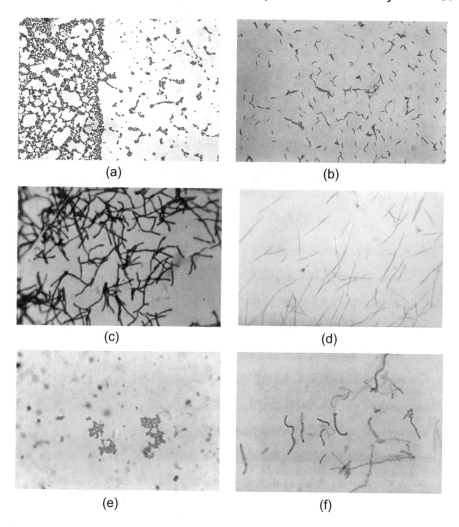

Figure 3.13   Microscopic images of bacteria frequently present in the human mouth
(a) Gram stain of *Streptococcus mutans* (X1000). Gram-positive cocci can be seen, many
of which are in pairs, short chains or groups
Dr. Richard Facklam, Centers for Disease Control and Prevention, USA
(b) Gram stain of a *Corynebacterium* species showing Gram-positive bacilli.   •
Dr. W.A. Clark, Centers for Disease Control and Prevention, USA
(c) Gram stain of an *Actinomyces* species showing Gram-positive bacilli (X1200)
Dr. Lucille K. Georg, Centers for Disease Control and Prevention, USA
(d) Gram stain of a *Fusobacterium* species showing long, slender Gram-negative bacilli
Dr. V. R. Dowell, Jr, Centers for Disease Control and Prevention, USA
(e) Gram stain of a *Veillonella* species showing Gram-negative cocci (X1125)
Gilda Jones, Centers for Disease Control and Prevention, USA
(f) Gram stain of a *Prevotella* species showing Gram-negative bacilli (X956)
Dr. Holdeman, Centers for Disease Control and Prevention, USA

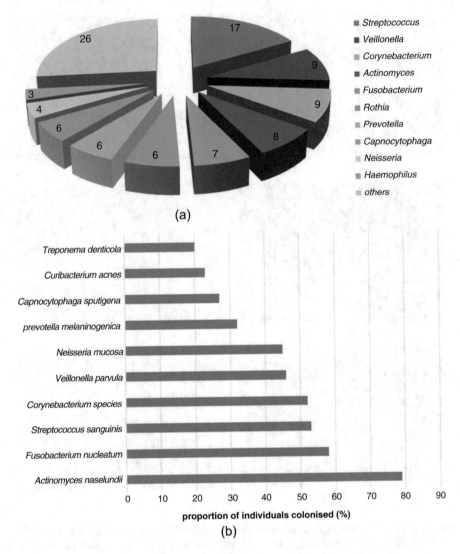

(a)

(b)

Figure 3.14 Bacteria found in dental plaque
(a) The relative proportions (%) of bacteria in dental plaque. Species belonging to the genera *Streptococcus, Veillonella, Corynebacterium, Actinomyces* and *Fusobacterium* are usually the predominant bacteria present. However, in the "others" category there are small proportions of a wide range of other bacteria
(b) Frequency of detection (%) of some of the bacteria that are usually present in dental plaque

Box 3.4 How are microbes named and classified?

Microbes, like all organisms, are named using the binomial system (i.e. the name is comprised of two words) and their classification is based on the traditional hierarchical system that biologists use to categorise all life on earth (Figure).

The first term of the organism's name refers to the genus to which it belongs while the second denotes the species – both terms are always italicised. For example, *Staphylococcus aureus*. The genus is often abbreviated e.g. *S. aureus* or *Staph. aureus.* Species with similar characteristics are considered to belong to the same genus e.g. *Staphylococcus epidermidis* has many properties in common with *Staphylococcus aureus*. Genera (the plural of "genus") with similar characteristics are grouped into a family, similar families into an order, similar orders into a class and similar classes into a phylum etc. (Figure)

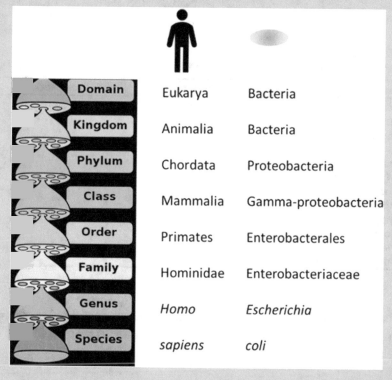

Domain	Eukarya	Bacteria
Kingdom	Animalia	Bacteria
Phylum	Chordata	Proteobacteria
Class	Mammalia	Gamma-proteobacteria
Order	Primates	Enterobacterales
Family	Hominidae	Enterobacteriaceae
Genus	*Homo*	*Escherichia*
Species	*sapiens*	*coli*

Figure. The hierarchy of biological classification and how it's applied to a human being (*Homo sapiens*) and a bacterium (*Escherichia coli*)
Pengo, Public domain, via Wikimedia Commons

Although the name of a microbe should always be written in italics, it's sometimes useful to use what are known as "common names" to refer to all of the species in a genus. For example, the word "staphylococci" can be used when talking about all the species within the genus *Staphylococcus*. The common name should never be italicised and only has a capital letter at the start of a sentence.

Box 3.4 (continued)

Often in describing the microbial communities present on humans or in the external environment it's convenient to use the higher orders of classification such as families or phyla. This is because such communities are often very complex and may consist of dozens of genera and, as most genera contain many species, thousands of species. For example, the genus *Staphylococcus* contains 45 different species. More than 20,000 different species have been detected in the human microbiota. To avoid confusing and overburdening the reader I will, in this book, usually avoid naming any individual species but will refer mainly to genera, families or phyla of microbes. This will make things much easier. Those interested can consult more detailed books and articles on the subject to find out more about particular species in a genus.

Table 3.2 Oral bacteria that can break down glycoproteins and/or proteins (+ = most species produce the enzyme; ? = data not available)

Organism	Sialidase	Glycosidase	Protease/peptidase
Actinomyces	+	+	+
Bifidobacterium	+	+	+
Capnocytophaga	+	+	+
Eubacterium	+	+	+
Haemophilus	?	+	+
Porphyromonas	+	+	+
Prevotella	+	+	+
Propionibacterium	+	?	+
Streptococcus	+	+	+
Treponema	+	?	+
Tannerella	+	+	+
Veillonella	?	+	?

These fungi are able to hydrolyse many of the macromolecules present in human tissues (Table 3.3) and some produce cytotoxins that kill human cells thereby releasing nutrients from them.

3.1.8.2 Microbes That Live in Our Stomach

The stomach is an aerobic environment with a temperature of 37 °C. It contains large quantities of acidic gastric juices whose function is to digest the chewed food that enters from the mouth. It's therefore a very acidic environment and only microbes capable of surviving at such a low pH (about 1.5) are found there. A wide range of nutrients are present in gastric juice and in

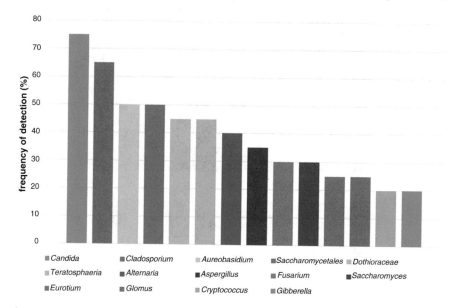

Figure 3.15 Frequency of detection of fungi in the mouths of healthy adults *PLoS Pathog.* 2010 Jan 8;6(1):e1000713. Characterization of the oral fungal microbiome (mycobiome) in healthy individuals

the food that arrives in the stomach. The microbes most frequently detected in the stomach (Figure 3.18a) are streptococci and bacteria belonging to the genera *Neisseria* (which are aerobes – Figure 3.17a), *Micrococcus* (Figure 3.17b) and *Veillonella*.

Streptococci tend to be present in the highest proportion (Figure 3.18b). The ability of streptococci and *Veillonella* species to break down macromolecules has already been described. Micrococci produce proteases, lipases and nucleases.

3.1.8.3   Microbes That Live in Our Small Intestine

The small intestine consists of the duodenum, ileum and jejunum (Figure 3.19) and is the region of the GIT in which most of our food is digested and where most of the nutrients in our food are absorbed into our bloodstream. Digested food passes rapidly through the small intestine and this, together with its low pH, makes it difficult for microbes to colonise this region. Nevertheless, microbial communities are found here and their composition tends to be complex. The main genera present (Figure 3.20) are *Streptococcus, Lactobacillus*

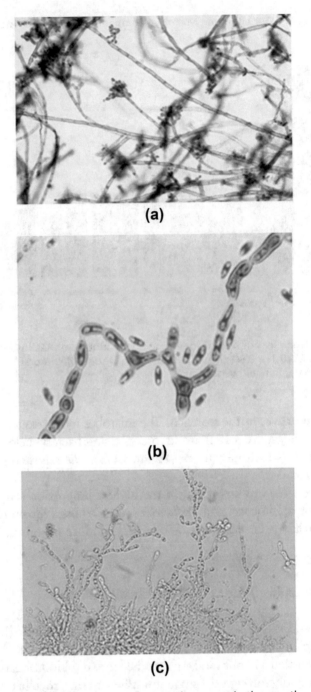

Figure 3.16 Images of fungi that are frequently present in the mouth
(a) Photomicrograph of a *Cladosporium* species (X400). The organism grows as multi-cellular filaments with a diameter of 2–6 μm and a length of 40-300 μm
Image courtesy of Dr. Libero Ajello, Centers for Disease Control and Prevention, USA
(b) Photomicrograph of an *Aureobasidium* species (X970). The organism grows as mul-ticellular filaments

Table 3.3 Hydrolytic enzymes and cytotoxins produced by fungi frequently present in the mouth

Enzyme	Candida	Cladosporium	Aureobasidium	Saccharomycetales
Protease	+	+	+	+
Lipase	+	+	+	+
Deoxyribonuclease	+	–	+	+
Polysaccharidase	–	+	+	+
Cytotoxin	+	+	+	–

(anaerobes), *Bacteroides* (anaerobes), *Prevotella* (anaerobes), *Veillonella* (anaerobes), *Fusobacterium* (anaerobes) and *Clostridium* (anaerobes).

3.1.8.4   Microbes That Live in Our Large Intestine

The large intestine consists of three main regions - caecum, colon and rectum (Figure 3.21). The material that arrives here contains constituents of our diet that haven't been digested in the small intestine. This material passes only slowly through the large intestine and this allows time for the establishment of microbial communities that consist of a huge number and variety of microbes. Indeed, in terms of numbers, most of the human microbiota resides in the large intestine. Each gram of the material present contains approximately 10^{11} bacteria and 10^9 viruses as well as an unknown number of archaea, fungi and protozoa. The environment of this region is very low in oxygen, has an almost neutral pH and has a huge supply of nutrients from our diet.

Although we can't digest many of the constituents of our diet (particularly the complex polysaccharides found in plants), the microbes that live in our large intestine can do this for us. It's here that we can clearly appreciate the important contribution our microbiota makes to our wellbeing. The material arriving in the large intestine consists of those macromolecules in food that the enzymes in our stomach and small intestine can't break down such as the polysaccharides xylan, pectin and cellulose. However, among the enormous variety of microbes that live in our large intestine, there are some that can convert these macromolecules to small compounds that are then absorbed by

---

Figure 3.16 (continued) Dr. Hardin, Centers for Disease Control and Prevention, USA (c) Photomicrograph of a *Saccharomyces* species belonging to the order *Saccharomycetales* (X400). The organism exists as oval cells that join together to form filaments.
Dr. Hardin, Centers for Disease Control and Prevention, USA

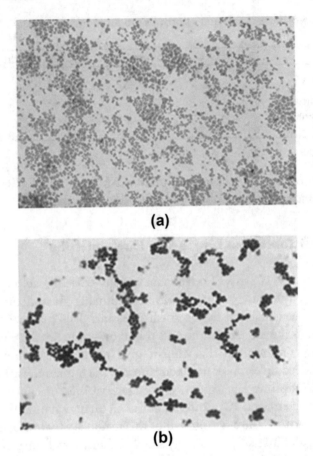

Figure 3.17 Images of bacteria frequently present in the stomach of humans
(a) Gram stain of a *Neisseria* species showing Gram-negative cocci, many of which are in pairs
Dr. W.A. Clark, Centers for Disease Control and Prevention, USA
(b) Gram stain of a *Micrococcus* species showing Gram-positive cocci, many of which are in bunches
Dr. Richard Facklam, Centers for Disease Control and Prevention, USA

the intestinal epithelium and used as nutrients and as an energy supply for our cells. It's been estimated that 10% of the energy needs of an adult are derived from compounds produced by bacteria in the colon. Not only do our intestinal microbes supply us with nutrients and energy but they also produce many of the vitamins that we need such as folic acid, isoprenoid, vitamin B6, vitamin B12, thiamine, biotin, riboflavin, pantothenic acid, nicotinic acid, pyridoxine and vitamin K. They also supply us with some of the amino acids that

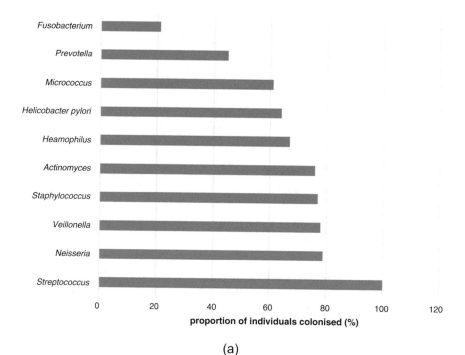

(a)

(b)

Figure 3.18   The gastric microbiota. The dominant bacteria are usually species belonging to the genera *Streptococcus, Prevotella* (anaerobes), *Micrococcus* (aerobes) and *Veillonella* (anaerobes)

(a) The frequency of detection of bacteria in the stomach of adults

(b) The relative proportions of bacteria present in the stomach of adults

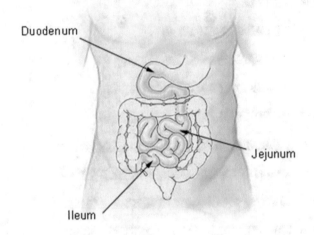

Figure 3.19 Diagram showing the three regions of the small intestine
This work is in the public domain in the United States because it is a work prepared by an officer or employee of the United States Government. https://doi.org/training.seer. cancer.gov/ss_module07_ugi/unit02_sec03_anatomy.html

we are unable to synthesise (these are known as "essential amino acids") – these include histidine, isoleucine, leucine and threonine.

Microbiologists are still at an early stage in trying to identify the microbes that live in the large intestine. Currently it's estimated that there are at least 15,000 different bacterial species, 1200 viruses, 270 fungi, 35 protozoa and 25 archaea. As for the total number of microbes present in the large intestine, the current estimate is that there are at least 10^{14}. It's been determined that the anaerobic bacteria in the large intestine are at least one thousand times more plentiful than aerobic species. As we'll see in Chapter 5, these anaerobes play an important role in the putrefaction of the human corpse. The variety of bacteria present is so large, and the differences between individuals so great, that it's difficult to say which, if any, are dominant. Nevertheless, *Bacteroides, Faecalibacterium, Bifidobacterium* and *Prevotella* (all of which are anaerobes) tend to be present in the highest proportions in many adults (Figure 3.22).

From Figure 3.22 we can see that some bacteria are invariably present e.g. *Bacteroides, Enterobacteriaceae* (facultative anaerobes), *Eubacterium* (anaerobes), *Bifidobacterium* (anaerobes), *Enterococcus* (facultative anaerobes) and *Clostridium* (anaerobes). What some of these bacteria look like is shown in Figure 3.23.

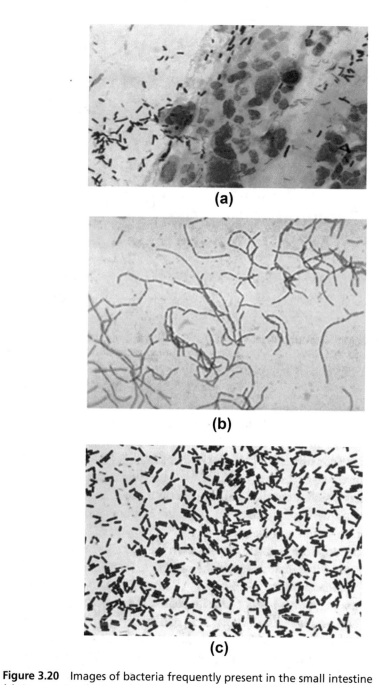

Figure 3.20 Images of bacteria frequently present in the small intestine
(a) Gram stain of a *Lactobacillus* species showing Gram-positive bacilli. Human epithe-lial cells (large red cells) and some unidentified Gram-negative bacilli are also visible
Joe Miller, Centers for Disease Control and Prevention, USA
(b) Gram stain of a *Bacteroides* species showing Gram-negative bacilli
Dr. V. R. Dowell, Jr, Centers for Disease Control and Prevention, USA
(c) Gram stain of a *Clostridium* species showing Gram-positive bacilli (X1000)
Don Stalons, Centers for Disease Control and Prevention, USA

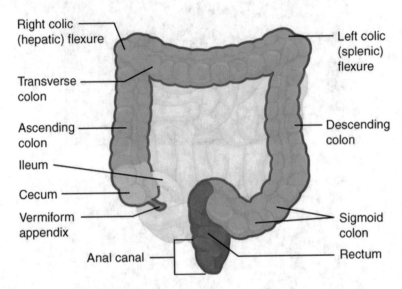

Figure 3.21 Diagram showing the main regions of the large intestine
OpenStax College, CC BY 3.0 <https://creativecommons.org/licenses/by/3.0>, via
Wikimedia Commons

As well as being able to break down macromolecules found in the human diet, many intestinal microbes can also hydrolyse the polysaccharides and glycoproteins that are constituents of human tissues (Table 3.4). More than 5000 different carbohydrate-degrading enzymes have been detected in the intestinal microbiota of which more than half can act on those carbohydrates found in humans. Consequently, intestinal microbes are important contributors to the decomposition of the human body.

Many intestinal microbes can also break down the proteins found in human cells and tissues and these include species belonging to the genera *Clostridium, Bacteroides, Parabacteroides, Alistipes, Fusobacterium, Prevotella, Propionibacterium, Enterococcus, Escherichia, Streptococcus, Bacillus, Staphylococcus* and *Lactobacillus* (Table 3.5). Collectively, these are able to degrade all of the proteins found in human tissues. However, only a few intestinal bacteria (mainly members of the genus *Bacillus*) are able to break down keratin, the protein found in hair and nails, so that this persists for longer than most human proteins. In contrast to bacteria, many fungi can degrade keratin and these include some that are present on the skin (*Malassezia*), on mucosal surfaces (*Candida*) and in the intestinal tract (*Aspergillus, Cladosporium* and *Fusarium* – Figure 3.24).

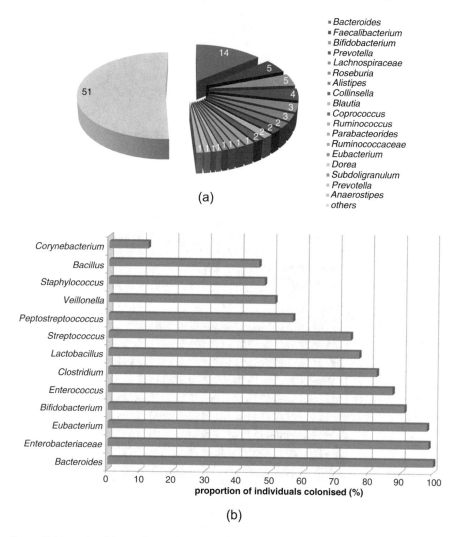

Figure 3.22   Microbiota of the large intestine
(a). Relative proportions (%) of the various genera in the large intestine of adults
(b) Frequency of detection of bacteria in the large intestine of adults. Figures represent mean values of the proportions (%) of individuals colonized

DNA is degraded by many intestinal microbes including members of the genera *Faecalibacterium, Bacteroides, Roseburia, Clostridium, Fusobacterium, Peptostreptococcus, Prevotella* and *Enterococcus*.

Fungi are also present in the colon and the most frequently detected genera are *Saccharomyces, Candida* and *Cladosporium* (Figure 3.25). These are able to hydrolyse a wide range of tissue macromolecules as described previously.

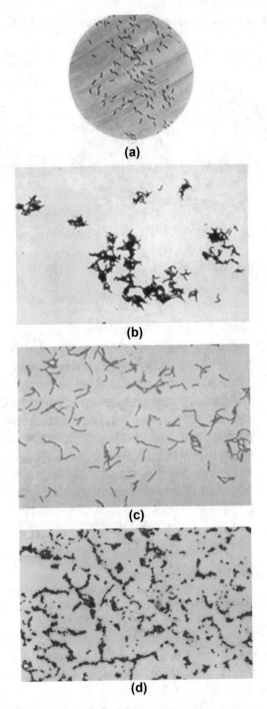

Figure 3.23 Images of bacteria that are invariably present in the colon and usually comprise a high proportion of the microbial community that lives there
(a) Gram stain of *Escherichia coli*, a member of the *Enterobacteriaceae*, showing Gram-negative bacilli
Centers for Disease Control and Prevention, USA

Table 3.4 Polysaccharide and glycoprotein degradation by representative genera of the gastrointestinal microbiota. (+ = some species are able to degrade the macromolecule; - = most species are unable to degrade the macromolecules;? = data not available)

Bacterial genus	Macromolecule present in human tissues			
	mucin	heparan	chondroitin sulphate	hyaluronan
Bacteroides	+	+	+	+
Bifidobacterium	+	-	-	?
Eubacterium.	-	-	-	?
Ruminococcus	+	-	-	?
Clostridium	+	-	+	+
Enterococcus	?	+	+	+
Lactobacillus	?	+	+	?
Akkermansia	+	?	?	?

Table 3.5 Breakdown of tissue proteins by genera of the gastrointestinal microbiota. (+ = some species are able to degrade the protein;? = data not available)

Microbial genus	Protein found in human tissues				
	Collagen	Fibronectin	Vitronectin	Elastin	Laminin
Bacteroides	+	?	?	+	?
Clostridium	+	+	+	?	?
Porphyromonas	+	+	?	?	+
Fusobacterium	+	+	?	+	?
Streptococcus	+	+	+	+	?
Bacillus	+	?	?	?	?
Pseudomonas	+	+	?	+	+
Staphylococcus	+	+	+	+	?
Prevotella	+	+	?	+	?
Lactobacillus	?	+	?	?	?
Enterococcus	+	?	?	+	?
Treponema	+	+	?	?	+
Candida	+	+	?	?	+

The gastrointestinal tract is home to a huge number and enormous variety of microbes. Collectively they are able to degrade all of the proteins, glycoproteins, polysaccharides, nucleic acids and fats that are present in humans. Furthermore, the vast majority of these microbes are able to carry out their

Figure 3.23 (continued) (b) Gram stain of a *Eubacterium* species showing Gram-positive bacilli
Don Stalons, Centers for Disease Control and Prevention, USA
(c) Gram stain of a *Bifidobacterium* species showing Gram-positive bacilli
Image courtesy of Bobby Strong, Centers for Disease Control and Prevention, USA
(d) Gram stain of an *Enterococcus* species showing Gram-positive cocci
Dr. Richard Facklam, Centers for Disease Control and Prevention, USA

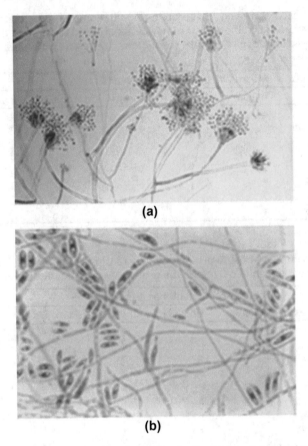

(a)

(b)

Figure 3.24 Keratin-degrading fungi that live in the human intestinal tract
(a) *Aspergillus* species (X 475). The photomicrograph shows the long, filamentous hyphae that are characteristic of fungi and also the spherical reproductive bodies known as conidia
Dr. Lucille K. Georg, Centers for Disease Control and Prevention, USA
(b) *Fusarium* species (X970). This shows the filamentous hyphae and oblong reproductive bodies known as microconidia
Dr. Hilliard F. Hardin, Centers for Disease Control and Prevention, USA

activities, grow and reproduce in the absence of oxygen. They are, therefore, ideally suited to degrade the human body even when oxygen levels in the corpse, whether above or below ground, have been depleted.

3.1.9   Which Microbes Live on My Skin?

The skin provides a mainly dry and acidic environment for the growth of microbes and the main nutrients available to them are lipids and proteins. These nutrients come from two main sources – the skin itself and secretions

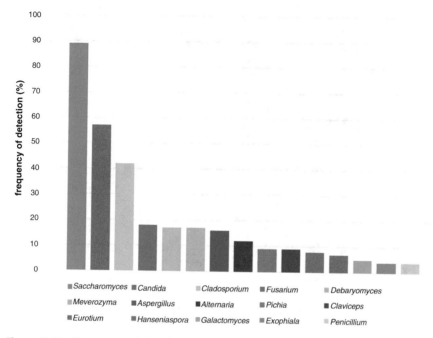

Figure 3.25 Frequency of detection of fungi in the colon of healthy adults

from the sweat and sebaceous glands (see Figure 2.2 in Chapter 2). The outer layer of the skin (stratum corneum – Figure 3.26), on which many microbes live, consists of several layers of dead skin cells (known as keratinocytes – Figure 3.27) which contain proteins (mainly keratin) and these are held together by lipid-rich material. The other main site of microbial colonisation is inside the hair follicle.

Most of the microbes that inhabit the skin are found on its surface (i.e. the stratum corneum) or in the hair follicles. As at other body sites, the microbial communities that inhabit skin are very complex and bacteria from 205 different genera have been detected. Nearly 5000 different species have been found on the hands alone. In addition to the wide range of bacteria detected, more than 130 different fungal species have been found as well as archaea and a variety of viruses. The main genera of microbes found on the skin (Figure 3.28) are *Corynebacterium, Staphylococcus, Propionibacterium, Micrococcus, Kocuria, Malassezia, Brevibacterium, Dermabacter* and *Acinetobacter*. Unfortunately, microbiologists have an awful habit of changing the names of the creatures they study and this can cause a lot of confusion. Recently, they decided that those species of *Propionibacterium* that are found on the skin should be given

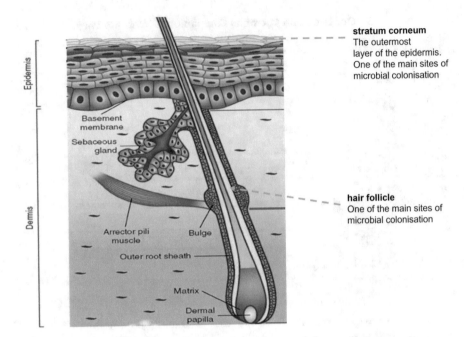

stratum corneum
The outermost
layer of the epidermis.
One of the main sites of
microbial colonisation

Basement
membrane

Sebaceous
gland

Arrector pili
muscle

Bulge

Outer root sheath

Matrix

Dermal
papilla

Epidermis

Dermis

hair follicle
One of the main sites of
microbial colonisation

Figure 3.26 Diagram of a cross-section through the skin showing the stratum corneum and the hair follicle which are the two main sites of microbial colonisation
Modification of Wong, D.J. and Chang, H.Y. Skin tissue engineering (March 31, 2009), StemBook, ed. The Stem Cell Research Community, StemBook, https://doi.org/10.3824/stembook.1.44.1, https://doi.org/stembook.org. / CC BY (https://doi.org/creativecommons.org/licenses/by/3.0)

Figure 3.27 Photomicrograph of the surface of the human skin stained with a red dye showing a regular pattern of adjacent keratinocytes which do not have a nucleus but contain large quantities of the protein keratin
From Piérard GE et al. (2014). *Scientific World Journal* 2014:462634. https://doi.org/10.1155/2014/462634. Published under CC BY 3.0

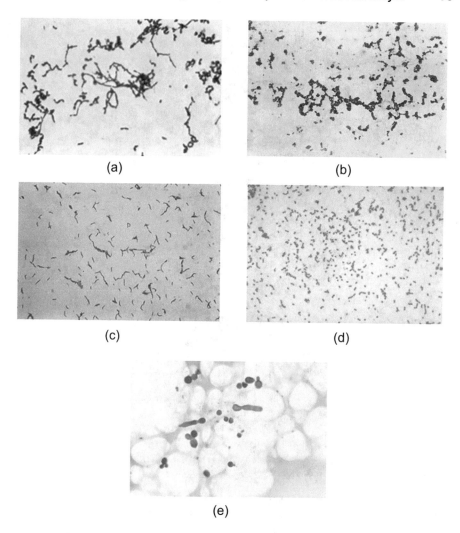

(a)

(b)

(c)

(d)

(e)

Figure 3.28 Images of the main genera of microbes found on the skin of healthy individuals

(a) Gram stain of a *Cutibacterium/Propionibacterium* species (X1150). This shows Gram-positive rod-shaped bacteria many of which have formed short chains
D. Lucille K. Georg, Centers for Disease Control and Prevention, USA

(b) Gram stain of a *Staphylococcus* species (X1250). This shows Gram-positive cocci which are mainly arranged in bunches
Dr. Richard Facklam, Centers for Disease Control and Prevention, USA

(c) Gram stain of a *Corynebacterium* species showing Gram-positive bacilli
Dr. W.A. Clark, Centers for Disease Control and Prevention, USA

(d) Gram stain of an *Acinetobacter* species showing Gram-negative cocci
Dr. W.A. Clark, Centers for Disease Control and Prevention, USA

(e) Photomicrograph of a *Malassezia* species. This is a fungus that can exist in two forms – small oval cells and short filaments
Dr. Lucille K. Georg, Centers for Disease Control and Prevention, USA

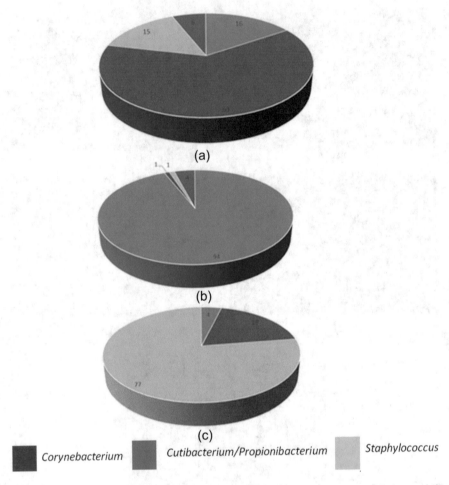

Figure 3.29 Relative proportions (%) of the main bacteria present in (a) the umbilicus – typical of moist skin sites, (b) on the forehead - typical of oily skin sites and (c). on the sole of the foot - typical of dry skin sites

the new name *Cutibacterium*. Unfortunately, many scientists still use the older name, so in this book it's best if we use both i.e. *Cutibacterium/Propionibacterium*).

The composition of the microbial community at a particular site on the skin is greatly influenced by the density of the sweat and sebaceous glands at that location. Depending on the density of these glands, three general types of skin site are recognised – oily (i.e. sebum-rich), moist and dry. The composition of the microbial communities at each of these types of site is different (Figure 3.29).

In general, *Cutibacterium/Propionibacterium* species (which are anaerobes) are the predominant organisms of sebum-rich regions such as the scalp, forehead, upper chest and back (Figure 3.29b). *Staphylococcus* species (facultative anaerobes) dominate at dry regions such as the arms, legs and the sole of the foot (Figure 3.29c). In moist regions (e.g. the armpits, groin, umbilicus), *Corynebacterium* species predominate (Figure 3.29a). As well as bacteria, fungi are found at many sites and these belong mainly to the genus *Malassezia* which are facultative anaerobes.

If we were to analyse the microbiota of a particular skin site (e.g. forehead or finger) of a group of adults and determine which genera of bacteria were present we'd find that the results would be very similar. However, if we took the analysis further and looked at which species were present and even determined the genetic profile of each bacterium present we'd find that each individual would have a collection of bacteria that was unique. This opens up the idea of each of us having a unique "microbial fingerprint" which could be used for identification purposes (Box 3.5). This also applies to the microbiotas found at all body sites.

Box 3.5 Forensic Microbiology

The DNA of every human being is different and we're now very familiar with the idea of DNA analysis being used to identify individuals during criminal investigations. Because the microbiota of an individual at a particular site is also unique to that individual there's a lot of interest in using these "microbial signatures" for identification purposes. Because DNA sequencing is so readily available, rapid and cheap, it's now possible to produce a DNA profile of the microbial community present at a body site in an individual.

As we go about our daily lives we touch a whole range of objects and each time we do so we leave behind a sample of our skin microbiota on those surfaces. It's now possible to take a sample from an object that's been touched and get a DNA profile of the microbial community that has been left on it (i.e. a "microbial fingerprint") and match this to a particular individual. A number of studies have shown that it's possible to identify individuals by DNA analysis of the microbial communities they have left behind on computer keyboards, computer mice, cell phones, various fabrics, doorknobs, bottles, pipes, books, drinking vessels, eyeglasses and steering wheels. While this is an interesting and promising approach there are a number of problems that need to be solved. First of all, an object that has been handled by more than one person will have microbes from each of these. Analysis of this mixture of microbial DNA won't provide a number of distinct "microbial fingerprints" but simply a mixture of them all. How to separate these out is very difficult. Another major difficulty is that we don't know how long microbial DNA last on a surface. Also, if the DNA of one of the microbes that was originally present in the skin community decays more rapidly than that of others then we might not be able to detect it. The resulting microbial fingerprint will not be a true reflection of what was present initially and could result in mis-dentification.

Microbial fingerprinting is an interesting example of the use of microbiology in medicolegal and criminal investigations, a branch of forensics known as "Forensic microbiology".

Table 3.6 Degradation of skin macromolecules by members of the skin microbiota

Macromolecule	Skin microbes able to break down the macromolecule	End products of breakdown
Lipids	*Cutibacterium/Propionibacterium, Staphylococcus, Corynebacterium, Malassezia, Acinetobacter*	Fatty acids, glycerol
Proteins (keratin, collagen)	*Cutibacterium/Propionibacterium, Staphylococcus, Corynebacterium, Malassezia, Brevibacterium, Acinetobacter, Dermabacter*	Amino acids
DNA	*Cutibacterium/Propionibacterium, Staphylococcus, Brevibacterium, Corynebacterium, Dermabacter*	Deoxyribose, bases, phosphate, nucleosides, nucleotides
Hyaluronan	*Cutibacterium/Propionibacterium, Candida, Streptococcus, Peptostreptococcus, Corynebacterium*	sugars
Chondroitin sulphate	*Cutibacterium/Propionibacterium, Streptococcus, Peptostreptococcus, Bacillus*	sugars

The skin consists of a variety of macromolecules including proteins (mainly keratin and collagen), polysaccharides (e.g. hyaluronan), lipids, proteoglycans (e.g. chondroitin sulphate) and DNA. In addition it does, of course, have large numbers of cells with all of their constituents. Many skin microbes are able to break down the macromolecules present in skin and examples are given in Table 3.6.

Again, as is the case for the intestinal microbiota, the vast majority of skin microbes can grow and reproduce in the absence of oxygen.

With regard to fungi, *Malassezia* species are invariably present on all regions of the skin where they comprise as much as 80% of all the fungi found there. They form oval-shaped cells that produce long filaments. They are facultative anaerobes and can hydrolyse proteins and lipids.

3.1.10   Which Microbes Live in My Respiratory System?

The respiratory tract (Figure 3.30) consists of two main regions: (i) the upper respiratory tract - the nose, pharynx and larynx and (ii) the lower respiratory tract - the trachea, bronchi, bronchioles and alveoli. Because all of its regions are open to the atmosphere via the nostrils and the mouth, microbes can easily gain access and so are found throughout the respiratory tract. The upper respiratory tract is heavily colonised by microbes whereas the lower regions are generally less densely colonised.

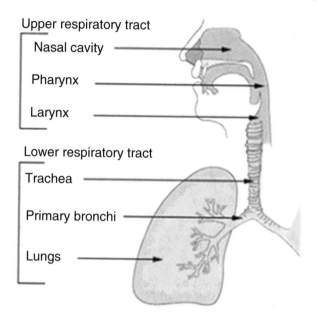

Figure 3.30 Diagram showing the main features of the human respiratory tract From https://doi.org/.training.seer.cancer.gov; funded by U.S. National Cancer Institute's Surveillance, Epidemiology and End Results (SEER) Program with Emory University, Atlanta, USA

The respiratory tract is predominantly an aerobic environment but anaerobic regions are present because of local anatomical features that hinder oxygen replenishment e.g. the convoluted surfaces of some epithelial cells and within the crypts of the tonsils. It's lined by a mucosa that is coated in a layer of mucous and this fluid is the main source of nutrients for respiratory microbes. Mucous consists mainly of water, mucins (which are complex glycoproteins), proteins, glycosaminoglycans and lipids. More than 250 different proteins have been identified in respiratory mucous.

The bacteria most frequently detected in the respiratory tract include species belonging to the genera *Streptococcus, Neisseria, Haemophilus, Moraxella, Staphylococcus, Corynebacterium, Propionibacterium, Prevotella, Porphyromonas,* and *Mycoplasma*. Images of most of these have been included previously, those that haven't are shown in Figure 3.31. Fungi (mainly *Candida* and *Malassezia*) and viruses are also present. Many members of the respiratory microbiota can degrade a variety of macromolecules present in human tissues and secretions (Table 3.7).

As at other body sites, the vast majority of members of the respiratory microbiota can grow and reproduce in the absence of oxygen.

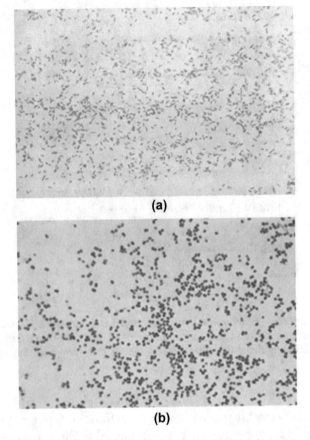

Figure 3.31 Images of bacteria frequently present in the respiratory tract
(a) Gram stain of a *Haemophilus* species showing Gram-negative short, rod-shaped bacteria
Dr. W.A. Clark, Centers for Disease Control and Prevention, USA
(b) Gram stain of a *Moraxella* species showing Gram-negative cocci, many of which are in pairs
Dr. W.A. Clark, Centers for Disease Control and Prevention, USA

The most frequently-encountered fungi in the respiratory tract are *Malassezia* species and these have been described above.

3.1.11 What About the Microbes That Live at Other Body Sites?

The genitourinary tracts of both men and women are also densely colonised by microbes but the total numbers are relatively small in comparison to those of the GIT, skin and respiratory tracts. Nevertheless, they have the ability to

Table 3.7 Breakdown of macromolecules by members of the respiratory microbiota

Macromolecule	Respiratory microbes able to hydrolyse the macromolecule	End products of hydrolysis
Lipids	*Staphylococcus, Cutibacterium/Propionibacterium, Corynebacterium, Moraxella*	Fatty acids, glycerol
Proteins	*Staphylococcus, Haemophilus, Neisseria, Micrococcus, Streptococcus, Porphyromonas, Prevotella, Cutibacterium/Propionibacterium*	Amino acids
Mucins	*Streptococcus, Micrococcus, Prevotella, Porphyromonas, Cutibacterium/Propionibacterium, Bacteroides.*	Amino acids, sugars
Hyaluronan	*Streptococcus, Cutibacterium/Propionibacterium, Staphylococcus, Bacteroides, Candida*	sugars
Chondroitin	*Streptococcus, Cutibacterium/Propionibacterium, Peptostreptococcus, Bacillus*	sugars
DNA	*Moraxella, Streptococcus, Staphylococcus, Cutibacterium/Propionibacterium, Corynebacterium*	Sugars, nucleotides, bases

hydrolyse a wide range of macromolecules found in human tissues and so will contribute to the decomposition of a human corpse.

3.2   How Do We Manage to Co-Exist With Our Microbiota?

From an early age we're told that microbes are very dangerous because they cause all sorts of awful, deadly diseases. The current COVID-19 pandemic due to severe acute respiratory syndrome coronavirus 2 has shown just how devastating a disease outbreak can be. We worry a lot about these disease-causing microbes as we learn that they lurk in the air, soil, in toilets, in dirt – everywhere. So, it then comes as a surprise when we find out that they actually live on us and are good for us (Box 3.6). So, how do we manage to live harmoniously, for most of our lives, with our microbiota? Our bodies have a variety of methods of defending us against microbes and these can be broadly classified into two main groups – innate and acquired immunity. Innate immune defences are a collection of physical and chemical systems that are in continual operation and are effective against all microbes. The following are its main components:

Box 3.6 The good side of our microbiota

The main benefits we gain from being colonised by so many microbes are the following:

1. The microbial communities that live on the various surfaces of our body (skin, mouth, respiratory tract etc.) are complex and stable due to a complex network of interactions among them. Because of this, it's very difficult for other microbes to become part of this network and so microbes from the environment are excluded. Some of these are highly pathogenic e.g. *Bacillus anthracis, Campylobacter jejuni*. This ability to repel environmental microbes is known as "colonisation resistance" and is very important in protecting us from harmful microbes. When we take certain types of antibiotics (particularly those that can kill a wide variety of microbes e.g. tetracyclines) our microbiota may be disrupted and this opens up the possibility of being infected by harmful species.

2. Many of the constituents of our diet such as complex plant polysaccharides (e.g. xylan, pectin, inulin) and proteins can't be broken down by the enzymes our digestive system produces. However, these macromolecules can be hydrolysed by microbes in the colon and the small molecules that are produced are absorbed into our bloodstream. It's been estimated that this process supplied as much as 10% of our daily energy needs.

3. Microbes in the intestinal tract produce a range of vitamins that are essential to us such as biotin, vitamin K, nicotinic acid, folate, riboflavin, pyridoxine, vitamin B12 and thiamine.

4. Our microbial symbionts play an important role in the development of the mucosal lining of our gut.

5. It's increasingly being realised that our indigenous microbiota plays a very important role in the development and maturation of our immune system.

6. Microbes in our gut can detoxify some of the potentially dangerous compounds present in our diet. These include substances such as methyl mercury and the cancer-inducing heterocyclic aromatic amines found in cooked meat.

- The skin and mucosal surfaces (i.e. the moist inner surfaces of our body such as the mouth, respiratory tract, gut, urethra, vagina etc.) act as a physical barrier that prevents microbes getting into our sterile, inner tissues.
- the cells of the skin and mucosal surfaces produce more than 400 antimicrobial compounds which accumulate on these surfaces and are able to kill many microbes
- the cells that make up the skin and mucosal surfaces are continually being shed (and replaced by new ones) and this means that any microbes that have attached to them are continually being removed as well
- Fluids such as urine, saliva and the mucus that covers mucosal surfaces continually flush away microbes from our internal body surfaces and deposit them into the stomach (in the case of saliva and respiratory mucus)

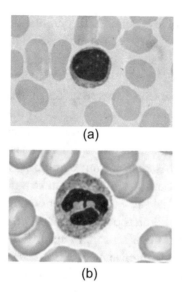

(a)

(b)

Figure 3.32 Some of the cells that are involved in the acquired immune response and so protect us from dangerous microbes
(a) Lymphocyte surrounded by red blood cells. This type of cell produces antibodies that bind to invading microbes
Guy Waterval / Apache License 2.0 (http://www.apache.org/licenses/LICENSE-2.0)
(b) A phagocytic white blood cell (known as a polymorph or neutrophil) that eats microbes, it's shown surrounded by red blood cells
Guy Waterval / Apache License 2.0 (http://www.apache.org/licenses/LICENSE-2.0)

where they are killed by the acidic gastric fluid or into the environment (in the case of urine).

Acquired immune defence systems operate in response to particular microbes and so only become active when we are exposed to such microbes. They may take days to weeks to become effective and involve particular types of cells (known as lymphocytes – Figure 3.32a) as well as specific proteins (known as antibodies). An acquired immune response is very specific and targets only one particular microbe, this is unlike the innate immune response which is effective against a huge range of microbes. The antibodies produced by lymphocytes bind to their target microbe which makes it easy to recognise by our phagocytic (i.e. cell-eating) cells (Figure 3.32b) which can then engulf and destroy the antibody-coated microbe. Vaccination against a particular pathogen primes the acquired immune system so that the lymphocytes and antibodies are produced more rapidly than usual.

So, given this impressive array of antimicrobial defence systems, how is it that our skins and mucosal surfaces are still covered in microbes? Shouldn't we be free of all microbes? The short answer to this question is that we don't really know. Since the time of Pasteur, microbiologists have tended to focus their attention on microbes that cause disease rather than on those that live happily with us – our symbionts. However, during the last few decades we've begun to appreciate the importance of our microbial symbionts and, consequently, have learnt far more about them. Our epithelial cells appear to be able to distinguish between microbes that are our symbionts and those that aren't. In other words it's as if we can recognise, and make a distinction between, what we could term a "microbial self" (our indigenous microbiota) and a "microbial non-self" (pathogenic microbes). Although we certainly don't know the whole story, it looks as though sensors in epithelial cells (known as Toll-like receptors – TLRs) can distinguish between our symbionts and pathogenic microbes. A set of 10 different TLRs have been detected so far and when these recognise that a particular microbe isn't one of our symbionts (and therefore could be harmful) then they signal to our immune system that it should get ready to mount a defensive response. On the other hand, if the TLRs recognise that a microbe is a member of our microbiota then no such signal is sent and it's left alone. Another possibility is that members of our indigenous microbiota are able to dampen down our immune system in some way so enabling them to co-exist with us. As research into our indigenous microbiota progresses, we will learn more about how we manage to live in harmony with the large number, and variety, of microbes that comprise our microbiota.

Although these immune defence systems are active and very effective while we are alive, once we die they stop operating. It's then that our long-term microbial companions turn against us and start to "eat the hand that fed them". How they do this will be described in Chapter 5.

3.3   Want to Know More?

The human microbiota in health and disease: an ecological and community-based approach. Michael Wilson. 2018. CRC Press.

Bacteriology of humans: an ecological perspective. Michael Wilson. 2008. Wiley-Blackwell
The healthy human microbiome. Lloyd-Price J, Abu-Ali G, Huttenhower C. *Genome Medicine* 2016 Apr 27;8(1):51. https://doi.org/10.1186/s13073-016-0307-y.

The Human Microbiome Project, National Institutes of Health, USA https://web.archive.org/web/20101210013519/http://nihroadmap.nih.gov/hmp/

The human microbiota in health and disease. Wang B, Yao M, Lv L, Ling Z, Li L. *Engineering* 2017; 3; 71-82
https://www.sciencedirect.com/science/article/pii/S2095809917301492

The Human Microbiome. BBC Science Focus
https://doi.org/sciencefocus.com/the-human-body/human-microbiome/

Normal Human Microbiota. Microbe Notes, Online Microbiology and Biology Study Notes
https://doi.org/microbenotes.com/normal-human-microbiota/

Introduction to the human gut microbiota. Thursby E, Juge N. *Biochemical Journal* 2017; 474 (11): 1823–1836. https://doi.org/10.1042/BCJ20160510
https://doi.org/portlandpress.com/biochemj/article/474/11/1823/49429/Introduction-to-the-human-gut-microbiota

The human gut microbiota; overview and analysis of the current scientific knowledge and possible impact on healthcare and well-being. European Commission, 2018
https://publications.jrc.ec.europa.eu/repository/bitstream/JRC112042/human_gut_microbiota_online.pdf

Role of the gut microbiota in nutrition and health. Valdes AM, Walter J, Segal E, Spector TD. *British Medical Journal.* 2018; 361 doi: https://doi.org/10.1136/bmj.k2179
https://www.bmj.com/content/361/bmj.k2179

The human microbiome and its impacts on health. Ogunrinola GA *et al. International Journal of Microbiology* 2020; Article ID 8045646
https://www.hindawi.com/journals/ijmicro/2020/8045646/

Managing the microbiome: how the gut influences development and disease. Weinstein N, Garten B, Vainer J, Minaya D, Czaja K. *Nutrients.* 2020 Dec 29;13(1):E74. https://doi.org/10.3390/nu13010074. PMID: 33383647
https://doi.org/.mdpi.com/2072-6643/13/1/74

Introduction to host microbiome symbiosis in health and disease. Malard F, Dore J, Gaugler B, Mohty M. *Mucosal Immunology* 2020 Dec 9:1-8. https://doi.org/10.1038/s41385-020-00365-4
https://www.ncbi.nlm.nih.gov/pmc/articles/PMC7724625/

The human respiratory system and its microbiome at a glimpse. Santacroce L *et al. Biology (Basel).* 2020 Oct 1;9(10):318. https://doi.org/10.3390/biology9100318

https://www.ncbi.nlm.nih.gov/pmc/articles/PMC7599718/

Role of the microbiome in human development. Dominguez-Bello MG, Godoy-Vitorino F, Knight R, Blaser MJ. *Gut* 2019;68:1108-1114 https://doi.org/gut.bmj.com/content/68/6/1108.full

Skin microbiome and its interplay with the environment. Callewaert C, Ravard Helffer K, Lebaron P. *American Journal of Clinical Dermatology* 2020 Sep;21(Suppl   1):4-11.   https://doi.org/10.1007/s40257-020-00551-x. PMID: 32910439
https://doi.org/ncbi.nlm.nih.gov/pmc/articles/PMC7584520/

The human microbiome and its impacts on health. Ogunrinola GA, Oyewale JO, Oshamika OO, Olasehinde GI. *International Journal of Microbiology* 2020; 2020: 8045646.
https://www.ncbi.nlm.nih.gov/pmc/articles/PMC7306068/

Impact of the human microbiome in forensic sciences: a systematic review. García MG, Pérez-Cárceles MD, Osuna E, Legaz I. *Applied and Environmental Microbiology* 2020 Oct 28;86(22):e01451-20. https://doi.org/10.1128/ AEM.01451-20. Print 2020 Oct 28

Challenges in human skin microbial profiling for forensic science: a review. Neckovic A, A H van Oorschot R, Szkuta B, Durdle A. *Genes (Basel).* 2020 Aug 28;11(9):1015. https://doi.org/10.3390/genes11091015
https://www.ncbi.nlm.nih.gov/pmc/articles/PMC7564248/

Trick or treating in forensics - the challenge of the saliva microbiome: a narrative review. D'Angiolella G, Tozzo P, Gino S, Caenazzo L. *Microorganisms.* 2020;8(10):1501. https://doi.org/10.3390/microorganisms8101501
https://www.ncbi.nlm.nih.gov/pmc/articles/PMC7599466/

4

Not a Pretty Picture – Our Appearance After Death

In this chapter you're going to find out how your body changes as it decomposes. These changes are certainly not pleasant to look at and therefore I'm not going to include any images of decomposing human corpses. However, if you want to see what a corpse looks like as it decomposes, I've provided information on where you'll find images relating to each of the stages of decomposition. General information on human corpse decomposition can be found on the websites of various research centres such as the ones listed in Box 4.1. Although photos of human corpses won't be included, the decomposition process will be illustrated using photos of other mammals and, for those with a strong stomach, these can be viewed in Appendix III at the end of the book.

The study of what happens to the human body after death is known as Taphonomy. This word comes from the Greek words "taphos" and "nomos" which mean burial and law respectively. Overall the process involves the conversion of organic matter (our tissues) to inorganic matter (minerals). Another way of looking at it is that it involves our passage from the biosphere (i.e. that zone of our planet that supports life) to the lithosphere (i.e. the rigid, rocky outer layer of Earth). Shakespeare, of course, had something to say about it and in "The Tempest" talks about the transformation that a human being undergoes if they drown:

> Full fathom five thy father lies;
> Of his bones are coral made;
> Those are pearls that were his eyes:
> Nothing of him that doth fade
> But doth suffer a sea-change

© The Author(s), under exclusive license to Springer Nature Switzerland AG 2022
M. Wilson, *Life After Death: What Happens to Your Body After You Die?*,
Springer Praxis Books, https://doi.org/10.1007/978-3-030-83036-6_4

Into something rich and strange.
Sea-nymphs hourly ring his knell:
Ding-dong.
Hark! now I hear them, ding-dong, bell.

Box 4.1 Research centres that study human corpses and their decomposition

(a) Forensic Anthropology Centre, University of Tennessee, Knoxville, USA

This centre was established by Dr. William M. Bass in 1987 in order to carry out research into human decomposition. One of the main resources at the center is the "Anthropology Research Facility", also known as "The Body Farm". This is an outdoor laboratory in which corpses are studied as they are left to decompose on various sites within the 2 acres of land owned by the Centre. It has a body donation program that enables people to donate their bodies for research. The centre also has a collection of more than 1700 skeletons and a Forensic Anthropology databank with details of more than 3,400 cases.

The website of the center is https://fac.utk.edu/

(b). The Forensic Osteology Research Station at Western Carolina University in the Blue Ridge Mountains was founded in 2007. As well as studying the decay of human corpses, it also trains dogs to detect human remains.

https://www.wcu.edu/learn/departments-schools-colleges/cas/social-sciences/anthsoc/foranth/forensic-anthro-facilities.aspx

(c). The Southeast Texas Applied Forensic Science (STAFS) Facility was established in 2009. It consists of four research divisions; the Outdoor Research laboratory, the Anthropology laboratory, the STAFS Skeletal Collection, and the Soft Tissue and Skeletal Trauma Laboratory. It carries out anatomical studies of human bodies and the ways in which the information obtained can be used by the medico-legal and science communities. Researchers at the center use bodies that have been donated to them for the purpose of scientific research.

Its website is: https://www.shsu.edu/centers/stafs/

(d). The first of such centres to open outside the USA is the Australian Facility for Taphonomic Experimental Research. This was opened near the city of Sydney in 2016.

https://www.uts.edu.au/about/faculty-science/after-facility/about-us

---

4.1 How Can We Find Out What Happens to a Body After Death?

Before we go any further, it's worth giving this question some thought as it'll give you an idea of why there are so many gaps in our knowledge of what actually happens to a human corpse as it decomposes. So, how would you go about studying this process? Ideally, you'd examine a corpse before it's been buried and take photos of it and a whole range of measurements (dimensions, weight, temperature etc.) as well as samples for microbiological and chemical analysis. Then, after burial, you'd dig it up and repeat all the above observations, tests and sampling. Then you'd re-bury it and repeat the whole process on several occasions. Ideally, you'd want to do this every day and over a period of several months or even years. That's a tremendous amount of work. But think for a minute. This will give you a lot of data but does it really tell you what usually happens to a corpse? The answer is no, because the act of digging up the body (exhumation) each day and then re-burying it will disrupt the normal processes that would otherwise occur. Each time you dig up the corpse you're disturbing it because this involves moving it (eventually, it will start falling to pieces) and it will be increasingly difficult to put it back exactly as it was. You're also changing the environment that surrounds it. Once you've dug it up you'll be exposing the corpse (and the surrounding soil) to oxygen in the atmosphere and this wouldn't usually happen. As you'll see in Chapter 5, oxygen has a dramatic effect on the microbial communities involved in decomposition. Repeated exposure to fresh sources of oxygen each time the body had been dug up would disrupt the normal course of events. Exposure of the soil that's been dug up not only affects its oxygen content but also its temperature, humidity, its compactness and the insect communities that live there. So, eventually, when you analyse all the data you've obtained, what this will show you will be the effects of repeated exhumation of a corpse – not what happens to a corpse under normal conditions.

Most of the studies of human corpse decomposition that have been carried out haven't used the "repeated exhumation" approach described above. Researchers have, instead, generally investigated corpses that have been left undisturbed out in the open air in research institutes. This isn't ideal as the corpses are unburied and, therefore, exposed to the atmosphere where they'll be subjected to conditions very different from those that would exist underground. This is a compromise and isn't ideal but at least it doesn't involve the repeated disturbances and environmental changes associated with the exhumation approach. It also allows us to observe, examine, and sample the corpse as frequently as we want, and without disturbing it. This approach also gives us a greater understanding of what happens to the corpses of the millions of mammals (and other animals) on the planet that die naturally and aren't buried.

Another problem for researchers is statistics. It's not scientifically acceptable to draw general conclusions based on the results of looking at a single corpse. For a start, you'll need to look at a large number of corpses. Exactly how many can be worked out by statisticians and this will depend on the type of observations you'll be carrying out. But, for the sake of argument, let's say the statisticians tell you that you need to look at 10 corpses. But, of course, each corpse is different. It's likely (and , in fact, it's known) that the decomposition process will be affected by gender, age, body size, body weight, health status at the time of death, cause of death etc. So, in order to carry out a scientifically and statistically valid investigation, you're going to have to study hundreds of corpses. You'll need an army of investigators and who's going to pay for them all? Then there's the question of ethics. Not many people like the idea of their body being studied in this way and this puts a severe limitation on the number of corpses available for study.

It's important to appreciate the problems that have been outlined above because they explain why we know so little about how our body decomposes after death. Such difficulties are also the reason why many of the published studies on the decomposition of human corpses have involved such small numbers of individuals – this will become particularly apparent in subsequent chapters.

4.2 The Decomposition of a Human Corpse Follows a Predictable Pattern

Extensive studies of human corpses (buried and unburied) have shown that decomposition generally occurs in a series of recognisable stages (Figure 4.1). Interestingly, a similar pattern is followed during the decomposition of other dead mammals such as pigs and mice. Studies of the stages involved, and their timings, are important aspects of Forensic Science i.e. the application of science to matters of criminal and civil law. Establishing the period of time that's elapsed since death (known as the "post mortem interval" - PMI) is crucial for establishing when a person died and this is particularly important in murder investigations (Box 4.4).

Figure 4.1 gives the approximate timings of each stage, however, these vary considerably and depend on factors such as the cause of death, body size, age at death, gender, whether or not the corpse is buried, whether or not it's been embalmed and the type of environment in which it's been placed. Also, a large number of environmental factors affect the rate of human corpse decomposition and these include the prevailing weather conditions, temperature, humidity, pH (i.e. the acidity or alkalinity) and the oxygen level. Nevertheless, if the corpse is left undisturbed, decomposition begins approximately four minutes after a person dies and then passes through several stages. Different investigators have identified between 4 and 9 stages, although most studies appear to agree on there being five main stages. However, although the concept of

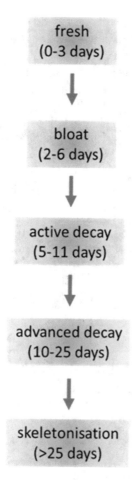

Figure 4.1   The main stages identified during the decomposition of a human corpse. The timing of each stage is only approximate as this is affected by many factors as described later.

discrete stages simplifies discussion of the whole process, it's important to bear in mind that the decomposition process is actually a continuum, rather than a series of distinct stages. Also, different parts of the body may go through the various stages at dissimilar rates which means that one part of the body may be in an early stage while other regions are at a more advanced stage. Let's now take a look at these various stages in more detail. As pointed out previously, most of what we know about the decomposition of a human corpse comes from studies involving corpses left out in the open air. However, we do know that corpses that are buried pass through the same stages, although the length of each stage is usually very different.

Box 4.2 Reversing decomposition - putting it all back together again

One of the most well known attempts to create a living human from dead parts was that featured in a book written by Mary Shelley (Figure a) in 1818 when she was 20 years old. The full title of this book is "Frankenstein; or The Modern Prometheus", but it's more popularly referred to as "Frankenstein".

Figure (a) A marble sculpture of Mary Shelley (1797–1851) by Camillo Pistrucci Purchase, Wrightsman Fellows Gifts, 2019, The Metropolitan Museum of Art. Public Domain.

Prometheus was a Greek god who created humans out of clay and also gave them the gift of fire. For this he was banished from Mount Olympus, the home of the gods, and tied to a rock where an eagle came every day to eat his liver. His liver was regenerated each night. Shelley's story involves the creation of a living man from body parts by a scientist – electricity provides the vital spark to animate the body of the creature (Figure b). Although most people think that the creature's name is Frankenstein, this is actually the name of his creator – Victor Frankenstein. The creature himself is not named in the book.

Box 4.2 (continued)

Figure (b) One of the earliest portrayals of Frankenstein's monster. Mr. T. P. Cooke, of the Theatre Royal, Covent Garden, in the character of the monster in the in the 1823 production of "Presumption; or, the Fate of Frankenstein"
Nathaniel Whittock, Public domain, via Wikimedia Commons

More recently, Bob Dylan (Figure c) has written a song about the creation of a new human being from body parts. This amazing song is "My own version of you" and is on the album "Rough and Rowdy Ways". If you haven't heard it then you must do so immediately on YouTube https://www.youtube.com/watch?v=pPLBbWNMDDk

Because of copyright restrictions, I can't reproduce the song lyrics here but he talks about having to go round lots of morgues and monasteries to collect all the various body parts he'll need. Ultimately, like Frankenstein's monster, it will be energised by electricity.

Box 4.2 (continued)

Figure (c) Portrait of Bob Dylan by Fionn Wilson
http://www.fionnwilson.co.uk/

4.3   What Are the Various Stages Involved in Decomposition?

Now let's take a look at the various stages a corpse goes through as it decomposes.

4.3.1   The Fresh Stage – 0 to 3 Days After Death

The first stage of decomposition involves a process known as "cellular autolysis" (or self-digestion) and this begins immediately after death and lasts approximately 1–3 days. The word autolysis comes from the Greek "auto" (self) and "lysis" (splitting) and refers to the cells of the body undergoing self-destruction. Autolysis occurs because the blood has stopped circulating around the body and therefore no oxygen or nutrients are being delivered to the cells and their waste products, such as carbon dioxide and acids, are not being removed. We'll describe the autolysis process in more detail in Chapter 5. What the corpse of a pig looks like during this stage is shown in Figure 4.2 of Appendix III.

Because the heart isn't pumping blood around the body, the force of gravity makes blood drain from those regions of the body that are furthest from the ground (usually the front of the body if the body is lying on its back) and this results in the skin becoming pale in Caucasians. This condition is known as "pallor mortis" (from the Latin: pallor "paleness", mortis "of death") and usually occurs within the first 30 minutes after death. This is also accompanied by a cooling effect known as "algo mortis" from the Latin word "algor" (coldness) that continues until the corpse reaches the temperature of its environment. In temperate climates the temperature drops by about two degrees within the first 12 hours and then continues to decrease at a rate of about 1°C per hour.

After about 2-8 hours, stiffening of the limbs takes place – this is the well-known "rigor mortis" (from the Latin word "rigor" meaning stiffness) and can last for 1 to 4 days. Those of us who watch detective movies will be familiar with this phenomenon because it's often used by forensic experts to establish when a murder victim had been killed. It starts in the facial muscles and this results in tightening of the jaw which is evident within 2-3 hours. The process peaks after about 12 hours, begins to recede after 24 hours and often completely disappears after 36 hours, at which point the muscles then relax again. This phenomenon occurs because the muscles are no longer being supplied with an important chemical called adenosine triphosphate (ATP). ATP is the source of energy for cells and muscles and is needed to maintain the muscles in a relaxed state. The jaw and neck muscles are affected first followed by muscles in the upper limbs and then the lower limbs. Figure 4.3 (in Appendix III) shows rigor mortis in a cow.

Because the lungs aren't working, no more oxygen is being absorbed into the blood and this lack of oxygen makes its colour change from bright red to deep purple. This oxygen-depleted blood accumulates in those regions of the body that are closest to the ground and results in a reddish-purple colouration, rather like a bruise (Figure 4.4) that's usually apparent after about 2 hours. This discolouration is known as "livor mortis" (from the Latin "livor"

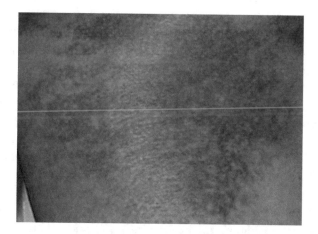

Figure 4.4   Bruised skin. Skin affected by livor mortis has a similar appearance. Modified from: Jonn Leffmann, CC BY 3.0 <https://creativecommons.org/licenses/by/3.0>, via Wikimedia Commons

meaning blueness) – it starts between 30 minutes and 2 hours after death and reaches a maximum after 8–12 hours. Any oxygen that does enter the body directly from the surrounding air is quickly used up by the bacteria and other microbes that are present in the corpse.

Flies and other insects usually arrive during this stage (discussed further in Chapter 6) although evidence of their activities doesn't usually become apparent until later during the bloat stage. They lay eggs from which larvae (maggots) will hatch and these will consume the flesh from the corpse.

4.3.2   The Bloat Stage – 2 to 6 Days After Death

Having consumed most of the oxygen present, the bacteria and other microbes (mainly from the gut and other parts of the body - see Chapters 3 and 5) that are present on and in the corpse start growing anaerobically (i.e. in the absence of oxygen). This results in the production of gases such as methane, hydrogen sulphide and ammonia which accumulate and begin to inflate the body. This usually starts in the face and lips and then the abdomen, giving the corpse a bloated appearance (this can be seen in Figure 4.5 in Appendix III) - the total volume of the corpse may actually double.

When microbes are growing anaerobically on a corpse (or on any other natural materials) they produce compounds that have an awful smell – this process is known as putrefaction. Putrefaction in a corpse is also accompanied by characteristic bloating, due to the gas formation already mentioned, and colour changes. A typical colour change involves the skin turning greenish-black due to the conversion of haemoglobin in the blood to sulfhaemoglobin by the gas hydrogen sulphide. This gas is produced by certain bacteria in the large intestine and the colour change tends to appear first on the skin of the lower abdomen. Putrefaction results in the complete loss of integrity of body tissues and the conversion of their constituent macromolecules into small molecules. The superficial veins of the skin also become visible as a pattern of lines that is usually called "marbling" (shown in Figure 4.6 in Appendix III)

The outermost layer of the skin (the stratum corneum - see Figures 2.3 and 3.26), consists of non-living cells and after death this readily becomes detached from the underlying tissues – this is termed "skin-slippage". When this occurs on the hands, it's described as "glove formation". Hair also begins to detach from the scalp.

Insect activity is pronounced at this stage and maggots are often visible on the head (shown in Figure 4.7 in Appendix III) and other sites. Although there may be lots of maggots on the outside of the corpse, there may be even larger populations inside it. Maggots often congregate in huge numbers (hundreds or thousands) to form what is known as a "maggot mass". The activities of all these microbes and maggots generates heat and often the temperature of the corpse increases during this stage. Ammonia is produced by insect larvae and this makes the surrounding soil alkaline which can result in the death of plants in the vicinity.

4.3.3 The Stage of Active Decay – 5 to 11 Days After Death

The start of this stage (shown in Figure 4.8 in Appendix III) is marked by the appearance of fluid from all of the main orifices - mouth, nose and anus. This fluid is forced out of the corpse because of the build-up of the pressure of the gases being produced inside the corpse. It's known as "purge fluid" and is red-brown in colour and foul-smelling.

The wall of the abdomen eventually bursts and the corpse, which has a wet and blackened appearance and an awful smell, now deflates as the internal gases are released. The compounds responsible for the foul smell are produced by microbes feeding on the tissues and include cadaverine, putrescine, skatole, indole and several sulphur-containing compounds. These compounds are responsible for attracting a lot of different insects to the corpse. During this stage most of the soft tissues are lost as they are broken down by enzymes from the corpse itself, from microbes and from insects. Adult insects and their larvae actively consume many tissues except for hair, bones and cartilage. The internal organs usually decompose in the following sequence: (a) the intestines and spleen, (b) the heart, kidneys, lungs and bladder and (c) the prostate or uterus. In general, maggots are very active during this stage. Flies can't consume the tougher parts of the body such as the skin, but beetles have chewing mouthparts and so can get to work on the skin. Several different beetle species can live in this way, or can eat the flies they find on the corpse. Other predatory insects, such as ants, wasps, mites and spiders, can feed on the insects (or their larvae) that are living on the corpse. Towards the end of this stage, the corpse begins to dry out and the main insects present tend to be scavenging and predatory beetles.

In certain circumstances, such as immersion of the corpse in water, or in a humid and warm environment, a process (known as saponification) may occur that results in the formation of a malodorous, whitish or grey waxy coating. This waxy material is known as grave wax or adipocere and is due to the decomposition of adipose tissue (see Chapter 2) by bacteria. The word adipocere comes from the Latin words "adip" and "cere" meaning fat and wax respectively. Some bacteria are able to hydrolyse the fats present in the adipose tissue to fatty acids (see Chapter 2) which then are converted to soapy materials. Adipocere formation is more common in infants, women and those who are overweight because their bodies contain a higher proportion of fat. The formation of this coating may inhibit further decomposition of the corpse and accounts for the fact that bodies immersed in cold water may be recovered virtually intact after several weeks.

4.3.4 Advanced Decay (or Post-Decay) Stage - 10 to 25 Days After Death

The nature and duration of this stage depends very much on the environment, particularly how moist it is. In general, the corpse dries out and little soft tissue is left and by the end of this stage only bones, dried skin and cartilage remain (as can be seen in Figure 4.9 in Appendix III). The remaining skin forms a leather-like sheet that clings to the bones. The rate of decay decreases and microbes start to ferment body fats producing butyric acid which has a cheesy smell. The foul smell associated with the corpse during the active decay stage usually disappears. Insect activities are usually greatly reduced.

4.3.5 Skeletal Decay Stage. >25 Days After Death

This stage is also often known as the dry stage or the skeletonization stage. At this point, all of the soft tissues have decomposed and all that remains are the dry skeleton and hair (shown in Figure 4.10 in Appendix III). Bone consists mainly of the mineral known as hydroxyapatite and collagen which is a protein (see Chapter 2). Further decomposition of these materials (known as "diagenesis") depends very much on the type of environment. For example, hydroxyapatite decomposes much more rapidly if the soil is acidic. Physical forces, such as the gnawing of bones by animals, will also speed up the process.

> **Box 4.3 Famous skeletons**
>
> The skeletons of many important historical figures have been identified, sometimes thousands of years after their death. Examples include King Tutankhamun of Egypt (1323 BCE), Xin Zhui of Hunan, China (3rd century BCE), Queen Eadgyth of Saxony (946 CE) and King Richard III of England (1487 CE). However, few skeletons have become famous for their appearance in a play. The exception to this is, of course, Yorik in Shakespeare's "Hamlet". In this play Hamlet comes across a gravedigger who is digging up some graves and removing the skeletons so as to provide room for more bodies. Hamlet points to a skull and asks who it is. The gravedigger tells Hamlet that it is the skull of Yorik. Hamlet remembers Yorik as the court jester who used to play with him when he was a child. Hamlet reflects on the transience of life and our ultimate fate in his short, but very famous, speech:

Box 4.3 (continued)

Alas, poor Yorick! I knew him, Horatio: a fellow of infinite jest, of most excellent fancy: he hath borne me on his back a thousand times; and now, how abhorred in my imagination it is! My gorge rims at it. Here hung those lips that I have kissed I know not how oft. Where be your gibes now? Your gambols? Your songs? Your flashes of merriment, that were wont to set the table on a roar? Not one now, to mock your own grinning? Quite chap-fallen? Now get you to my lady's chamber, and tell her, let her paint an inch thick, to this favour she must come; make her laugh at that.

Hamlet and Horatio before the Gravediggers. 1843. Eugène Delacroix
Rogers Fund, 1922, Metropolitan Museum of Art, Public Domain

Table 4.1 The characteristic features associated with the various stages in the decomposition of a human corpse

Stage of decomposition	Main features
Fresh	• corpse appears fresh but is decomposing internally due to autolysis and the activities of microbes • pallor mortis • algo mortis • rigor mortis • livor mortis • little insect activity
Bloat	• inflation of abdomen due to gases produced by microbes • putrefaction and smell generation • discolouration of skin • skin marbling • skin slippage • internal temperature increases • surrounding soil becomes alkaline • considerable insect activity
Active decay	• fluid seepage from mouth, nose and anus • bursting of abdomen • skin is moist and blackened • skin broken in several places • foul smells produced • considerable insect activity • loss of internal organs
Advanced decay	• fermentation of body fats • cheesy smell • corpse consist mainly of bones, hair and skin • insect activities greatly reduced.
Skeletonisation	• only bones and hair remain • bone starts to break down (diagenesis)

Table 4.1 summarises the main features associated with the various decomposition stages of a human corpse. Although much of what we know about these decomposition stages has come from studies of corpses left above ground, it's important to emphasise that buried corpses go through similar stages, the main difference being that they occur over a much longer time scale.

The various stages of decay of a human corpse have been depicted in a series of paintings by the Japanese artist Morishige Kinugasa in 1670 (Figure 4.11). This series, known as "The nine meditations on the impurity of the body", portrays the decomposition of the body of a beautiful young woman. Meditation on these paintings was meant to dispel desire.

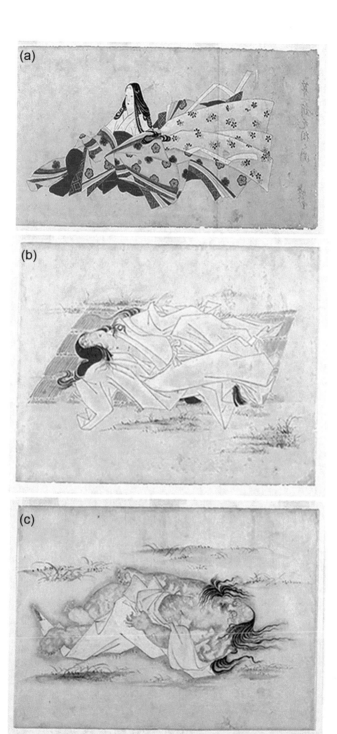

Figure 4.11 From "The nine meditations on the impurity of the body" by Morishige Kinugasa (1670 CE). (a). The lady while alive. (b) The lady has just died. (c) bloat stage (d) active decay stage (e). advanced decay stage (f) skeletonisation stage

Images courtesy of the Collectie Stad Antwerpen, MAS; Public Domain, CC0 1.0 Universal (CC0 1.0)

Figure 4.11   (continued)

4.4 The Smell of Death - "I think, I think, I smell a stink, it comes from y...o...u"

Decomposition of a corpse is accompanied by dramatic chemical changes and many of these result in powerful, usually unpleasant, smells due to the release of compounds that evaporate into the air – these are known as "volatile" compounds. These compounds are produced by microbial metabolism of the simple molecules (amino acids, fatty acids etc) that have been released by the enzymatic breakdown of macromolecules (see Chapter 5). While we may find these offensive, many animals think otherwise. Many insects, for example, are attracted by these smells because they signal to them the presence of suitable food. We'll see that the various chemicals emitted by a corpse at different stages of its decomposition are a cue for particular insect species and this is responsible, in part, for the succession of insects that feed on the corpse.

So, let's take a look at what's going on. Studies have shown that several hundred volatile compounds are released from a decaying corpse. The number of such compounds increases to a maximum during the active decay stage (Figure 4.12) and then decreases until none are detectable in the skeletonisation stage.

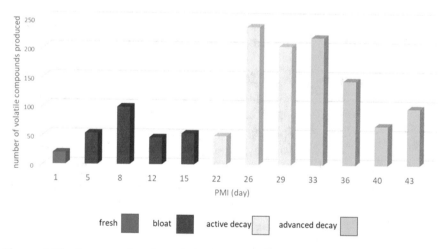

Figure 4.12 Compounds released during the decomposition of a pig left above ground in the open air
Enhanced characterization of the smell of death by comprehensive two-dimensional gas chromatography-time-of-flight mass spectrometry (GCxGC-TOFMS) Dekeirsschieter J et al. PLoS One. 2012;7(6):e39005. This is an open-access article distributed under the terms of the Creative Commons Attribution License, which permits unrestricted use, distribution, and reproduction in any medium, provided the original author and source are properly credited

Table 4.2 Nature of the volatile compounds released during the decay of a pig corpse

Decomposition stage	Main types of compounds produced
Fresh	Alkenes (a type of hydrocarbon)
Bloat	Alcohols, ketones, carboxylic acids, aldehydes, aromatic compounds
Active decay	Nitrogen-containing compounds, sulphur-containing compounds, carboxylic acids
Advanced decay	Ketones, alkanes, carboxylic acids, sulphur-containing compounds

Table 4.3 Chemicals that act as attractants for various insects

Chemical	Insects attracted
CO_2	Most insects
Ammonia, hydrogen sulphide and carboxylic acids	Many flies
Butyric acid and other carboxylic acids, ketones,	Many beetles
Butyric acid and 2-butanone (a ketone)	Dung beetles
Indole, Sulphur-Containing Compounds And Phenol	Blow flies
Dimethyl disulphide or trisulphide	*Calliphora vicina* (a blow fly), *Lucilia caesar* (greenbottle)

As well as the number of volatiles released varying with decomposition stage, the types of compound detected also vary and these are summarised in Table 4.2.

These volatile compounds are produced by the action of enzymes (human, microbial and insect) on tissue components. Proteins give rise to a variety of nitrogen-containing compounds (e.g. ammonia) and sulphur-containing compounds (e.g. hydrogen sulphide); carbohydrates produce mainly alcohols, ketones and carboxylic acids; while lipids produce mainly hydrocarbons, carboxylic acids, aldehydes and ketones. The metabolism of many compounds produces carbon dioxide (CO_2). Many of these volatiles can be sensed by insects which then hone in on their source i.e. they function as attractants (Table 4.3):

4.5   Factors That Affect the Rate of Decomposition

As mentioned previously, the time taken for a body to decompose varies greatly and skeleton formation can occur after a few days or could take many years. A variety of factors affect the decomposition rate and these include not only the characteristics of the body itself, but also the cause of death and the

Table 4.4 Factors that influence the rate of decomposition of the human body

Factors relating to the nature of the body	Environmental factors
Age at death	Buried or left above ground
Physique i.e. thin or fat	Presence or absence of clothing
Presence of wounds	Temperature
Cause of death	Humidity
Underlying diseases and infections	Soil type (if buried)
Use of certain medications	Depth of burial (if buried)
	Type of coffin material (if buried)
	Wind (can lower temperature and remove water)
	Immersion in water

environment in which the corpse has been left (Table 4.4). With regard to body characteristics, a thin body is turned into a skeleton more quickly than one that has more flesh. Because of their much smaller body mass, infants and juveniles skeletonise more rapidly than adults. Wounding, either before or after death, increases the rate of decomposition because it allows easier access to the internal tissues for insects and microbes. The corpse of someone who has died from a serious infection will decay rapidly because microbes will already have been distributed throughout their body while they were alive.

Environmental factors have a substantial effect on the rate at which a corpse decomposes. Consequently, uncertainty regarding these factors often makes it difficult to determine an accurate value of the PMI which is very important in forensic investigations (Box 4.4). Generally, a corpse left above ground decays much more rapidly than one that's been buried (Figure 4.13). From Figure 4.13 it can be seen that when a corpse is left above ground, half of its mass will be lost within 4 days whereas this takes about 17 days for one that's been buried. This is because the temperature and oxygen level below ground are generally lower and the soil that surrounds the corpse hinders its invasion by insects and other animals as well as restricting the escape of gases. The activities of microbes, insects and other animals are, therefore, greatly reduced below ground and so decomposition occurs much more slowly. Also, many types of insects are unable to gain access to buried corpses. The rate of decomposition of a corpse in different environments is governed by what is known as Casper's rule. This states that the degree of decomposition of a corpse that's been lying on the soil surface for 1 week corresponds to that of a corpse that's been in water for 2 weeks or buried in soil for 6–8 weeks. While a corpse left above ground may reach the skeletonization stage within a few weeks, for a buried corpse this usually takes between 3 and 12 years under favourable conditions. However, when conditions are not conducive to decay, then skeletonisation of a buried corpse can take hundreds, or even thousands, of years.

Box 4.4 Determining the postmortem interval (PMI)

Determining the period of time that has elapsed between the time a person died and the finding of their body (the postmortem interval, PMI) is an important aspect of forensic investigation. When the cause of death is suspicious and is the subject of criminal investigation, establishing the PMI and, consequently, the time of death can be central to establishing the innocence or guilt of a murder suspect.

A wide range of techniques are available for determining the PMI and which of these is used often depends on just how recently death has occurred. Some are only appropriate within hours of death while others can be used for corpses that are much older. In general, the sooner a corpse is examined after death then the greater the chances of an accurate PMI and time of death estimate. Conversely, the longer this time period then there is an increased likelihood of error in estimating the time of death.

The techniques used to estimate the PMI are based on observation and measurement of the extent of the changes that are known to occur following death. These include a fall in temperature (algor mortis), discolouration of the skin (livor mortis), muscle rigidity (rigor mortis), putrefaction, chemical composition of tissues and fluids and insect colonisation. However, the rates at which all of the above changes take place depend on a number of factors including the nature of the body (male, female, age etc), the cause of death and the environment (temperature, humidity etc) in which the corpse is found. These variables mean that determining an accurate value for the PMI is challenging.

Measurement of the core (i.e. rectal) temperature of the corpse is the most intensively investigated of the methods available and is the most precise and reliable. However, after the corpse has reached the same temperature as that of its environment this method is no longer of any use. Generally this approach is applicable only during the early postmortem period, up to about 24 hours. During the first 6 h after death, the margin of error is at least 2 hours and this increases to 3 hours during the next 14 hours. A margin of error of 4.5 hours can be expected during the following 10 hours. Other methods that are useful in establishing the PMI during this early postmortem period include determining the concentration of potassium ions or urea in the vitreous humour of the eye and determining the degree of electrical and/or mechanical excitability of various muscles

Identification of what insects are present on a corpse (and their stage of development – see Chapter 6) is useful for estimating longer PMIs – particularly when these are greater than 72 hours.

Other methods of determining PMIs are being actively investigated and include analysis of:

(i) the microbiota of various organs and/or regions of the corpse
(ii) the microbiota of the soil adjacent to the corpse
(iii) the chemical composition of the soil adjacent to the corpse
(iv) the extent of degradation of DNA or proteins in muscles and in blood
(v) the volatile compounds emitted by the corpse

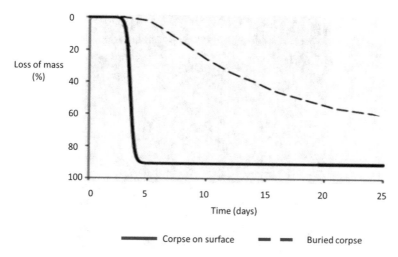

Figure 4.13 Loss of mass of a corpse when left on the surface and when buried

The nature of the surrounding soil has a significant effect on the decomposition of a buried corpse. In general, decomposition is slower in soils with a high content of clay than in sandy or loamy soils. Clays consist of smaller particles than sand or loam and this makes them less permeable so resulting in anaerobic conditions that decrease the activities of many soil microbes and insects. The type of coffin also affects the rate of decomposition. Decomposition takes place more rapidly in coffins made of pine or spruce than in those made from oak, zinc or lead.

4.6 Where's My Mummy?

Regardless of whether a corpse is left above or below ground, the rate of decomposition is faster in warmer climates whereas lower temperatures will slow the process down. This is mainly because of the effects that temperature has on the rate of metabolism and growth of microbes – these generally decrease by about 50% for each 10°C drop in temperature. Furthermore, at temperatures below 6°C most insect activity ceases. At very low temperatures (less than -5°C), decomposition may be halted completely as in the case of mammoths found in Siberia after 40,000 years (Figure 4.14). This is because temperature affects the growth rate of microbes as well as which types of insect are able to live on the corpse. A simple formula has been derived that can be used to estimate the time needed for the skeletonization of a corpse

Figure 4.14 The head of the 22,500 year old Yukagir mammoth. In 2002 the head of a woolly mammoth, (*Mammuthus primigenius*) was extracted from the permafrost near the village of Yukagir in Siberia. The head, without the trunk, was almost completely covered with skin. Other parts of the same animal were recovered later.
Stacy from Minneapolis, CC BY 2.0 <https://creativecommons.org/licenses/by/2.0>, via Wikimedia Commons

that has been left lying on the ground: y = 1285/x. In this formula, y is the number of days it takes to become skeletonized and x is the average temperature in °C during the decomposition process. As an example, if the average temperature of the environment is 10^0C, then the number of days required for skeletonization will be 1285/10 i.e. 128.5 days.

Mammoths aren't the only animals to have been preserved in very cold environments – the bodies of humans have also been recovered from frozen regions of the Earth. Perhaps the most well-known of these is Otzi, the iceman (Box 4.5).

Box 4.5 Otzi, the iceman

In 1991 the frozen body of a man was revealed after the glacier in which he had been encased had partially melted. The glacier was in the Tisenjoch pass in the Alps on the Italian side of the Italian/Austrian border. He was named Otzi after the place where he was found - the Ötztal Valley Alps. Later investigations

Box 4.5 (continued)

showed that the body was that of a 45-year-old man who had died approximately 5,300 years ago after having been wounded by an arrow – he also had a head injury. His skin (which had 61 tattoos), bones and other tissues were well preserved. Many of his clothes were also found including a coat made of animal skins, a fur hat, a grass cape and leather shoes. His belongings included a copper axe, flint dagger, bow, quiver of arrows and a leather pouch.

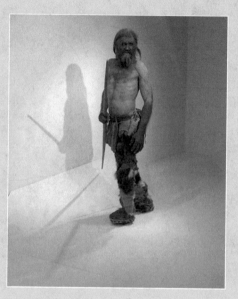

Figure. A reconstruction of Otzi
Andre, Schade, CC BY 3.0 <https://creativecommons.org/licenses/by/3.0>, via Wikimedia Commons

His body and belongings are housed in a museum devoted to him in Bolzano, Italy. Further information about the discovery and Otzi's life can be obtained from the website of the museum https://www.iceman.it/en/

A human corpse decomposes faster in a humid climate than in a dry one. This is because many microbes need a relative humidity of more than 60% to grow and reproduce. A dry and windy environment above ground can quickly remove water from a corpse which slows decomposition and can result in its preservation – a process known as mummification. A picture of a mummified Peruvian male is included in Appendix III (Figure 4.15).

As well as corpses being preserved naturally by low temperatures and dry conditions, the process of mummification can also be undertaken deliberately by humans (Box 4.6). This was practiced by many ancient civilisations including the Egyptians, Chinese, Incas and Guanches (the original inhabitants of the Canary Islands).

Box 4.6 Artificial Mummification

The ancient Egyptians are perhaps the most well known practitioners of mummification and the processes involved have been extensively researched. The oldest Egyptian mummies are approximately 5,500 years old. The main stages involved in mummification were the following:

(a). the body was washed in palm wine

(b) a cut was made in the left side of the body and the liver, lungs, stomach and intestines were removed, washed and placed in a mixture of salts (known as "natron") to dry them out. A metal hook was passed into the skull via the nose and the brain was pulverised and then pulled out through the nose. The inside of the body was washed in palm oil and spices. The heart (regarded as being the seat of thought and feeling) was not removed.

(c). the body was covered in, and stuffed with, natron to dry it out.

(d). the body was washed to remove the natron and covered in oils

(e). The dehydrated internal organs were wrapped in linen and placed back inside the body. Linen, leaves and sawdust could be placed inside the body to make it look more natural.

(f). the body was treated with oils, wrapped in linen and coated in resin.

A photograph of the mummy of Ramses II, an Egyptian pharaoh, is included in Appendix III.

The extent to which a corpse is exposed to the environment also affects decomposition. For example a naked corpse will decay more rapidly than one that's clothed or wrapped in blankets or plastic sheets. This is mainly because insects and environmental microbes won't be able to gain access so easily. A body submerged in water decays at a much slower rate because of the lower temperature, low oxygen level and reduced insect colonisation. Also, decay in the ground is slower when a body is buried in a sealed container or coffin.

The nature of the soil in which a corpse is buried also affects the rate of decay. Decomposition is more rapid in porous, permeable and light soils, because these allow a relatively free exchange of oxygen and water from the atmosphere as well as the escape of the gases produced during putrefaction. In general, the deeper the corpse is buried then the slower the decay rate because of the lower temperature, poorer access of oxygen and fewer insects.

Now that you've seen how the appearance of a corpse changes as it decomposes, let's take a look at how these changes are brought about by microbes (Chapter 5) and by insects (Chapter 6).

4.7   Want to Know More?

Forensic Anthropology Center, University of Tennessee, Knoxville, USA
https://fac.utk.edu/

Revolution in death sciences: body farms and taphonomics blooming. A review investigating the advantages, ethical and legal aspects in a Swiss context Varlet V *et al.*, *International Journal of Legal Medicine.* 2020 Sep;134(5): 1875-1895.
doi: 10.1007/s00414-020-02272-6. Epub 2020 May 21.

Evaluation of postmortem changes. Almulhim AM, Menezes RG.
https://www.ncbi.nlm.nih.gov/books/NBK554464/

Postmortem changes. French K, Jacques R.
PathologyOutlines.com website.
http://www.pathologyoutlines.com/topic/forensicspostmortem.html

Stages of decomposition. Australian Museum, Sydney
https://australian.museum/learn/science/stages-of-decomposition/

Decomposition. The Forensics Library
http://aboutforensics.co.uk/decomposition/

Forensic Anthropology. The Forensics Library
http://aboutforensics.co.uk/forensic-anthropology/

Deadly secrets - the science of decomposition. Forbes S, Blau S, Voss S.
Australian Academy of Science.
https://www.science.org.au/curious/decomposition

A website that provides educational resources on forensic medicine and pathology.
http://www.forensicmed.co.uk/

Post mortem interval
http://www.forensicmed.co.uk/pathology/post-mortem-interval/

The secret lives of cadavers: how lifeless bodies become life-saving tools. National Geographic.

https://www.nationalgeographic.com/news/2016/07/body-donation-cadavers-anatomy-medical-education/

Recent advances in forensic anthropology: decomposition research. Wescott DJ. *Forensic Sciences Research*, 2018; 3:4, 278-293, DOI:10.1080/2096179 0.2018.1488571
https://www.tandfonline.com/doi/pdf/10.1080/20961790.2018.1488571?n eedAccess=true

What is forensic science? BBC Science Focus Magazine
https://www.sciencefocus.com/the-human-body/what-is-forensic-science/

Challenges in human skin microbial profiling for forensic science: a review. Neckovic A *et al. Genes (Basel)*. 2020 Aug 28;11(9):1015. doi: 10.3390/genes11091015
https://www.ncbi.nlm.nih.gov/pmc/articles/PMC7564248/

Forensic applications of microbiomics: a review. Robinson JM *et al. Frontiers in Microbiology* 2021 Jan 13;11:608101. doi: 10.3389/fmicb.2020.608101. eCollection 2020
https://www.ncbi.nlm.nih.gov/pmc/articles/PMC7838326/

Early post-mortem changes and stages of decomposition in exposed cadavers. Lee Goff M. *Experimental and Applied Acarology* 2009; 49:21–36

Decay Process of a Cadaver. João Pinheiro. In "Forensic Anthropology and Medicine" Editors: Aurore Schmitt, Eugénia Cunha, João Pinheiro. Pages 85-116. Springer Nature, 2006

Decomposition of Human Remains. Robert C. Janaway, Steven L. Percival, Andrew S. Wilson. In: "Microbiology and Aging; Clinical Manifestations" Editor: Steven L. Percival. Pages 313-334. Springer Nature, 2009.

The effect of repeated physical disturbance on soft tissue decomposition—are taphonic studies an accurate reflection of decomposition? Adlam RE, Simmons T. *Journal of Forensic Sciences* 2007; 52(5):1007–1014

5

The Great Betrayal – Our Own Cells and Our Symbionts Turn Against Us

We've seen in Chapter 2 that the corpse of a human is a very rich resource because it offers a large quantity, and wide range, of nutrients to any creature that doesn't have an aversion to eating dead bodies. However, it's important to appreciate that the type and quantity of nutrients, as well as the nature of the environment, provided by a corpse change with time. In other words, it's a dynamic resource rather than a static one. This is due to two main factors. First of all, in most climates the body gradually loses water and dries out. Secondly the activities of microbes, insects and other animals result in the removal of some nutrients but the addition of others, in the form of the waste products of the organisms that are feeding on the corpse. Because of these environmental changes, the types of creature (both microscopic and macroscopic) present on the corpse gradually change – this process is known as "community succession" and is important in many ecological systems. In this chapter we're going to focus on how microbes make use of the nutrients present in a human corpse. The ways in which insects, the other main group of creatures involved in corpse decomposition, use the nutrients present in human tissues will be described in the next chapter.

© The Author(s), under exclusive license to Springer Nature Switzerland AG 2022   **117**
M. Wilson, *Life After Death: What Happens to Your Body After You Die?*,
Springer Praxis Books, https://doi.org/10.1007/978-3-030-83036-6_5

5.1   It All Starts With Autolysis – the Damage Is Self-Inflicted

But before we get to what our treacherous microbial symbionts do to our body, we first of all have to come to terms with the role played by our very own cells in kick-starting the decomposition process. In Chapters 1 and 4 we mentioned that within minutes of death, autolysis begins. This cellular self-digestion involves the release of a wide range of degradative enzymes that had previously been safely stored in membrane-enclosed vesicles (lysosomes and peroxisomes) inside cells (Figure 2.13). While they're trapped within these structures, they can't damage the cell in any way. In a living cell, lysosomes carry out a lot of important functions. For example, they digest viruses, bacteria and other microbes that have been taken in (i.e. phagocytosed) by the cell. They dispose of damaged organelles (e.g. mitochondria) and injured cells and break down proteins that have been absorbed by the cell. The enzymes involved are of a type known as hydrolases (a contraction of "hydrolysis" and "-ase") because the process requires water. These enzymes break down macromolecules into much smaller molecules, often down to their basic building blocks. They are classified (Box 2.3, Table 5.1) on the basis of the type of macromolecule they act on e.g. proteases, glycosidases, lipases and nucleases. The molecules produced are then used by the cell as a source of energy or are used to manufacture new macromolecules for a variety of purposes within the cell.

Autolysis occurs because the blood has stopped circulating and therefore no oxygen or nutrients are being delivered to the cells. Also, their waste products, such as carbon dioxide and acids, aren't being removed. The build-up of

Table 5.1 The main types of enzymes that are present in lysosomes

Enzyme	Target molecule(s)	Molecules produced
glycosidases (> 50 different types)	Polysaccharides and other carbohydrates	Oligosaccharides, monosaccharides and other simple sugars
Proteases (> 20 types)	proteins	Peptides, amino acids
Sulphatases (> 10 types)	proteoglycans, glycoproteins and glycolipids that contain sulphate groups	Sulphate, proteins, lipids and polysaccharides
Lipases	lipids	Glycerol and fatty acids
Phosphatases	Molecules that have phosphate groups attached	Phosphate
Nucleases	Deoxyribonucleic acid (DNA) and Ribonucleic acid (RNA)	Nucleotides

carbon dioxide results in an acidic environment and this causes all of the membranes surrounding, and within, each cell to rupture (Figure 5.1). The enzymes that were previously kept safely stored in lysosomes and peroxisomes are released and these start to break down the components of the cell. Lysosomal enzymes include proteases, glycosidases, lipases and nucleases (Table 5.1).

Peroxisomes contain at least 50 different enzymes many of which can oxidise compounds such as fatty acids, amino acids and uric acid. Hydrogen peroxide is produced during these reactions and this is a very powerful oxidising agent that can react with, and break down, many of the compounds present in the cell.

The release of this huge variety of enzymes into the cell and then their spilling out into the tissues means that all of the macromolecules in the cells and tissues will eventually be broken down to their low molecular mass building blocks (Figure 5.2 and Box 5.1). This process is usually referred to as "hydrolysis" because it requires water, as well as the appropriate enzyme. The small molecules produced (amino acids, sugars, glycerol, fatty acids, nucleotides etc.) will then be available as nutrients for the microbes and insects that are feeding on the corpse.

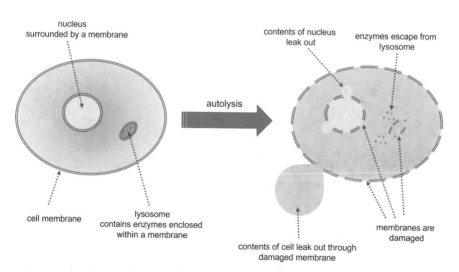

Figure 5.1 Autolysis of human cells occurs after death. The membranes inside, and surrounding, the cell break down

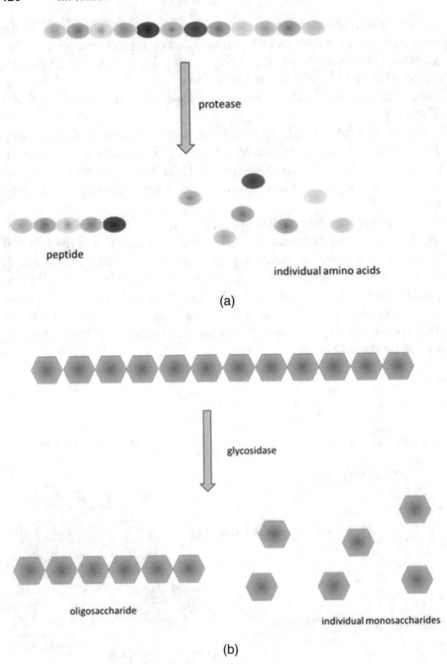

Figure 5.2 The enzymatic breakdown of macromolecules results in the liberation of low molecular mass compounds that can be used as nutrients by many creatures
(a) Protein hydrolysis
(b) Polysaccharide hydrolysis
(c) Lipid hydrolysis
(d) Nucleic acid hydrolysis

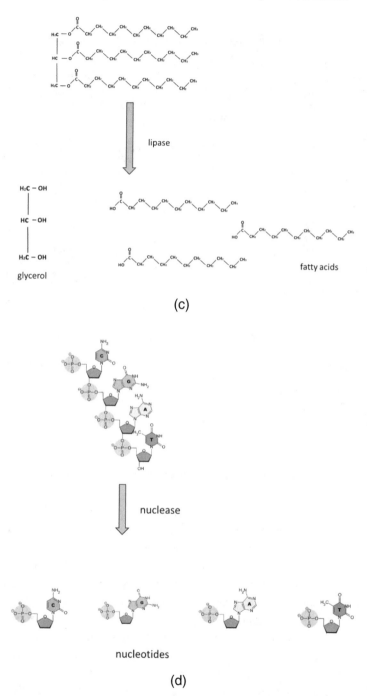

Figure 5.2 (continued)

Box 5.1 Degradation of macromolecules

Evidence of the degradation of macromolecules in human corpses has been obtained in a number of studies. The quantities of particular macromolecules in tissues can be determined by staining with dyes that bind preferentially to certain macromolecules and then determining how the intensity of staining varies with the PMI. In one such study, the dye safranin O, which stains proteoglycans, was used to determine the quantity of these macromolecules present in cartilage taken from the knees of corpses at different PMIs. As can be seen in the figure, the intensity of staining decreased with time showing that degradation of the proteoglycans had taken place.

Figure. Samples of cartilage taken from the knee of a human corpse 1, 12 and 36 days after death
Safranin O without fast green is the best staining method for testing the degradation of macromolecules in a cartilage extracellular matrix for the determination of the postmortem interval. Alibegović, A et al. Forensic Science Medicine and Pathology 16, 252–258 (2020). https://doi.org/10.1007/s12024-019-00208-0

5.2   And Now the Microbes Can Take Over

The presence of large quantities of sugars, amino acids and other small molecules in a human corpse, together with the release of more of these by autolysis, provides an excellent nutrient supply for microbes and insects (Figure 5.3a). The corpse will, initially, also provide plenty of water as well. Once this supply of readily-available nutrients has been used up by the microbes that previously kept us safe, these organisms can then use their own enzymes to continue the breakdown of the macromolecules present in cells and tissues and so the supply of nutrients will continue for some time (Figure 5.3b).

The small molecules (sugars, amino acids, fatty acids etc.) that are released by the breakdown of macromolecules are converted to a wide range of even smaller molecules when they're used as nutrients by microbes. Examples of these are listed in Table 5.2 and many of these have an important role to play in corpse decomposition because they are volatile and can act as attractants for various insects (see Chapter 6).

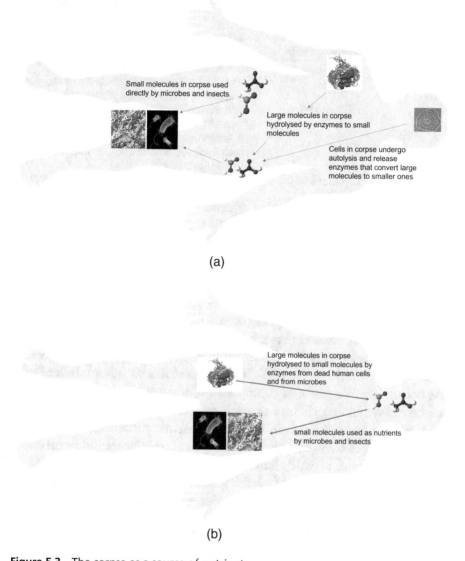

Small molecules in corpse used directly by microbes and insects

Large molecules in corpse hydrolysed by enzymes to small molecules

Cells in corpse undergo autolysis and release enzymes that convert large molecules to smaller ones

(a)

Large molecules in corpse hydrolysed to small molecules by enzymes from dead human cells and from microbes

small molecules used as nutrients by microbes and insects

(b)

Figure 5.3 The corpse as a source of nutrients

Acetic acid: Jynto and Ben Mills, Public domain, via Wikimedia Commons
Formic acid: Paginazero, Public domain, via Wikimedia Commons
Protein: Jawahar Swaminathan and MSD staff at the European Bioinformatics Institute, Public domain, via Wikimedia Commons
(a) Nutrients for microbes, insects and their larvae are provided by the small molecules already present in the corpse and those that are produced by the hydrolysis of macromolecules by enzymes released following the autolysis of cells
(b) Later, once all the small molecules initially present have been used up, the only source of further nutrients are the macromolecules present in tissues. These are broken down by enzymes present in the corpse and produced by microbes

Table 5.2 Compounds produced as a result of the microbial metabolism of the building blocks of macromolecules

Macromolecules	Building block of macromolecule	Compounds produced by microbial metabolism
Polysaccharide	Sugars	Carbon dioxide, water, carboxylic acids (acetic, butyric etc), alcohols (ethanol, methanol etc), ketones, aldehydes
Protein	Amino acids	Carbon dioxide, hydrogen sulphide, indole, skatole, methane, ammonia, putrescine, cadaverine, methanethiol, ethanethiol, dimethyl sulfide, dimethyl trisulfide
Lipid	Glycerol and fatty acids	Carbon dioxide, water, fatty acids, glycerol
NucleiAc acids	Nucleotides	Phosphate, ribose, deoxyribose, carbon dioxide, water, purines, pyrimidines, xanthine, uric acid, ammonia, acetic acid

5.3 It All Comes Down to Ecology

In order to understand what's going on, we first of all need to think about the nature of the environment that a fresh human corpse provides for microbes. At the time of death there'll be plenty of small molecules present in tissues, in blood vessels and within the GIT (from digested food). The autolysis of cells that occurs immediately after death will also liberate more of these. So, our treacherous symbionts will have no problem feasting on these easily-accessible nutrients. We have also seen in Chapter 3 that the communities inhabiting various body sites are well equipped with the enzymes needed to break down the macromolecules of which human tissues are composed. These provide an important additional source of nutrients once the initial supply of small molecules has been exhausted. While we are alive, the innate and acquired immune responses (described in Chapter 3) prevent our symbionts from damaging our tissues but, after death, these no longer function which means that this valuable source of nutrients can now be used by the microbes that are present. However, as we've seen in Chapter 3, microbes are found only on the surfaces (both inner and outer) of our bodies – our internal tissues are completely microbe-free. We can, therefore, expect some delay before microbes gain access to many of our tissues. This means that decay will progress at different rates in the various regions of the body.

While we are alive, the blood system delivers oxygen (as well as nutrients) to all parts of our body. Immediately after death, this system stops functioning, causing oxygen levels to fall in all tissues other than those (such as the skin) that are exposed to the atmosphere. So, at the start of decomposition aerobic microbes and facultative anaerobes (those that can live in the presence or absence of oxygen) will be able to grow on the corpse. However, oxygen consumption by these organisms will soon use up the available supply and we would expect the aerobes to gradually disappear. As we'll see later, this is what happens and anaerobes start to predominate and produce foul-smelling compounds. During the bloat stage, however, the gases produced by the microbes can rupture the skin and this will allow oxygen into the tissues from the atmosphere and provide an opportunity, though temporary, for the growth of aerobic microbes.

While we are alive, our internal tissues are maintained at a neutral pH (pH = 7.0). However, following death, the carbon dioxide that's produced by all human cells is no longer removed by the bloodstream and therefore accumulates and produces carbonic acid. Furthermore, the absence of oxygen changes the metabolism of the cell which starts to derive its energy from the fermentation of glucose and this produces lactic acid. These two processes result in a decrease in pH (i.e. the cells and tissues become acidic) which causes the cells to lyse as described above. Accumulation of lactic acid in the muscles after death also contributes to this acidification of the tissues. However, this doesn't present much of a problem to microbes. Within the vast assembly of different species that make up the human microbiota are many that can grow and reproduce over a wide range of pHs. Unlike the internal tissues, the skin and stomach are very acidic with a pH of 4-5 and 1.4 respectively. While we're alive, only microbes able to tolerate these extreme pHs will be able to survive in these regions. But, following death, these pH extremes gradually disappear because the underlying physiological processes that are responsible for them have stopped. This means that other microbes are now able to colonise and survive in these regions and take part in their decomposition.

Our bodies are maintained at a constant temperature of 37°C while we're alive and our microbial symbionts are very comfortable with this and grow best at this temperature. However, when we die our body temperature decreases (see Chapter 4) in a predictable fashion. Those who watch detective movies will know that the temperature of a corpse can be used to determine the likely time of death. Its final temperature will be that of the ambient temperature of the environment and this will, in non-tropical countries, vary greatly with the season. In temperate climates the final temperature, even during the summer, will fall well below the 37°C maintained by the body in life. This temperature decrease has a profound effect on our microbial symbionts. In general, the rate at which microbes grow decreases by about 50% with every 10°C fall in temperature so the microbes living on a corpse will start to grow far more slowly. In contrast, those microbes that live in the environment in which the corpse is placed are used to the temperature that prevails there. These environmental microbes, therefore, are at an advantage over our microbial symbionts from the point of view of temperature.

5.4 Ch...Ch....Ch....Changes (With Thanks to David Bowie)

A very important point to grasp at this stage in our analysis of what's going on is that the activities of microbes (and, later on, those of insects) bring about dramatic changes to the corpse so that its physical and chemical characteristics continually change. In other words, the corpse environment gradually changes. This brings about a change in the types of microbes and insects that can survive on the corpse – such changes are known as "community succession" (Figure 5.4). Imagine, for ease of discussion, that our corpse is in an air-tight container and the container has thick walls of steel so that nothing can get in or out and that the container is in a temperature-controlled room so that it neither warms up nor cools down. What will then take place is a series of changes known as "autogenic succession" – meaning that all changes are due to what goes on inside the container because nothing from the external world can affect it. "Autogenic" comes from two Greek words and means "produced from within". Those microbes inside the corpse that can use the nutrients that are present will grow and reproduce – let's consider two of these

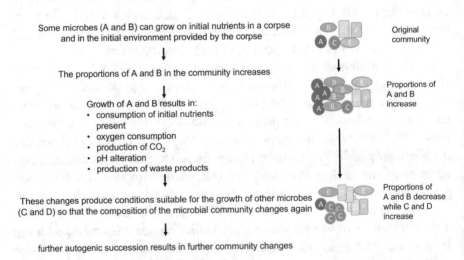

Figure 5.4 Autogenic succession that occurs in a microbial community inhabiting a corpse

and call them A and B. Other microbes present (let's call them C and D) that can't use these nutrients won't be able to grow. But after a while A and B will have used up the nutrients that they were so fond of and so their growth will slow down. At the same time, A and B will have produced waste products (e.g. alcohols, acids, carbon dioxide and other gases) that C and D may be able to use as nutrients and so these will start to grow and reproduce and come to dominate the community while the proportions of A and B will decrease. C and D will, in turn, deplete the nutrients that they are fond of and will produce different waste products that other microbes (E and F) can use. So the proportions of C and D will decrease while E and F will increase. And so on.

But these types of changes are not just happening with respect to the nutrients that are present in the corpse. The same applies to the oxygen and CO_2 levels, and also to the pH (the acidity/alkalinity). Consequently, changes in these factors that are brought about by microbial activities will also affect what microbes can grow and become dominant.

Finally, an important issue that affects the composition of any microbial community is inter-microbe competition. So far we've assumed that the different microbes in these communities ignore one another but this is far from the truth. Microbes don't all live together in harmony – they compete for scarce resources and many have developed weaponry to enable them to dispose of their competitors (Box 5.2).

Box 5.2 Inter-microbial warfare

The microbes present in any location (known as a "habitat") on our planet must compete for the resources that are available within that habitat. Consequently they have evolved a variety of strategies to fight off competitors to enable them to gain the upper hand and get unique access to those resources. A tactic commonly-used by bacteria is to produce proteins that can kill, or at least inhibit the growth of, other bacterial species. Such proteins are known as bacteriocins and are very effective antibacterial compounds. More than 400 different bacteriocins have been identified and there's considerable interest in the possibility of using them to treat human infections.

Some bacteria produce enzymes that can kill other bacterial species – examples include lysozyme, lysostaphin and many proteases.

Many microbes produce small molecules that can kill other microbes and these include hydrogen peroxide and a whole range of acids such as acetic, formic, lactic, succinic and propionic acids.

The above description gives you an idea of what's happening to the microbial communities that live on a corpse. However, you've got to remember that, as described in Chapter 2, each part of the human body has a different microbial community at the time of death so the actual changes that occur at each body site will be different. We can't really talk about the "corpse microbiota" i.e. the microbial community of a corpse – we have to think about the communities at different sites in and on the corpse (Box 5.3). Finally, remember that the situation in the real world is more complicated than what's been described above because a corpse is not an isolated system and the exchange of materials between it and its environment is possible and does occur. We also have to take into account that the temperature of the corpse will be gradually decreasing (at least in temperate climates) and will be drying out – these changes will affect the types of microbes (and insects) that can flourish. The take home message is that corpse decomposition is a dynamic process (Figure 5.5) with continuous changes taking place – these are best understood in terms of ecology.

Box 5.3 What's in a name?

The microbial community that lives in a particular habitat on planet Earth is known as the microbiota or microbiome of that habitat. These terms can, of course, be used to refer to the microbes that live on us (the human microbiota/microbiome) or a specific part of us e.g. the lung microbiota/microbiome. When it comes to our body after death we can refer to the microbes that live on the corpse as the corpse microbiota/microbiome or the necrobiome/necrobiota. Recently two new terms have been proposed to distinguish between the microbial communities that live on the internal organs of a corpse (the Thanatomicrobiome) and those that live on its external surfaces (the Epinecroticmicrobiome).

5.5   Mass Migration – A World (or, At Least, a Corpse) Without Borders

In Chapter 3 we saw that we can predict with a fair degree of certainty the nature of the microbial community that's likely to be present at a particular body site while a person is alive. Another important lesson from Chapter 3 is that these microbial communities are very different from one another. For example, the mouth is dominated by facultative anaerobes such as streptococci, in the large intestine the dominant microbes are anaerobes (such as *Bacteroides*) while the skin is dominated by a mixture of facultative anaerobes (such as staphylococci) and anaerobes (such as *Cutibacterium*). However,

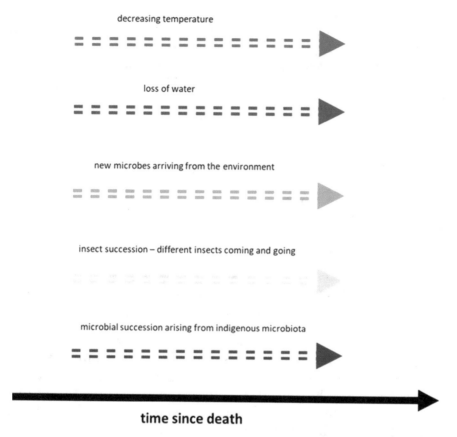

decreasing temperature

loss of water

new microbes arriving from the environment

insect succession – different insects coming and going

microbial succession arising from indigenous microbiota

time since death

Figure 5.5 The dynamic nature of corpse decomposition. In addition to these changes to the corpse itself, the environment in which the corpse has been placed is also fluctuating

death changes all this and distinctions between microbial communities gradually disappear. The autolysis that occurs soon after death results in a decrease in the integrity of our tissues – they start falling to pieces. In other words, the original demarcations that existed between tissues become less apparent. As an example, think of the large intestine. This is, basically, a tubular structure filled with digested food and microbes and it's separated from the surrounding connective tissues (that are completely microbe-free) by a thin layer of epithelial tissue i.e. the intestinal wall. Autolysis results in disruption of the epithelial wall which gradually disintegrates. This means that enormous numbers of intestinal microbes can now move across into the surrounding connective tissues which are exciting new regions offering a wide range of nutrients (Figure 5.6). It's the equivalent of humanity's discovery of the Antarctic.

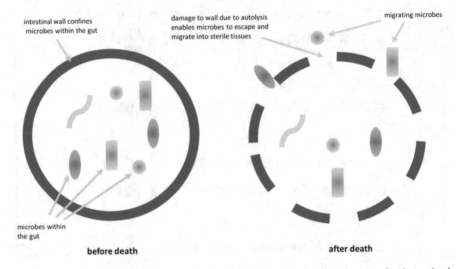

intestinal wall confines
microbes within the gut

damage to wall due to autolysis
enables microbes to escape and
migrate into sterile tissues

migrating microbes

microbes within
the gut

before death   after death

Figure 5.6 The autolysis that occurs after death results in damage to the intestinal wall which enables microbes to escape and colonise the sterile tissues that surround the intestinal tract

Similarly, the microbes inhabiting the respiratory tract can cross the newly-disrupted mucosal layer that separates them from the extensive network of blood vessels and connective tissue that surround the lungs. Skin microbes can penetrate the epithelium and then wander freely into the underlying connective tissue. The surface area of the large intestine is about 2 m² (similar to that of the skin) while that of the whole of the intestinal tract is 32 m². The respiratory mucosa has a surface area of 160 m². These figures give you an idea of just how large these mucosal surfaces are – imagine the enormous numbers of microbes that will quickly get access through these to our internal tissues after death. For a short time following death, a few of these migrating microbes might be gobbled up by the occasional white blood cell that has survived. But soon no more white blood cells, antibodies or other components of the body's immune system, will be produced. Our defences are down and out for the count. There's nothing to stop these massive microbial invasions - we're very quickly going to be constituents of the microbial world.

5.6   Knock, Knock – Who's There?

Trying to establish which microbes are involved in the decomposition of a human corpse isn't easy. Immediately after death, the microbial communities present at a particular body site start off the decomposition process at that site. However, this soon changes because of the microbial migrations described above that allow microbes from one body site to spread into other sites. So, in order to investigate the microbiology of corpse decomposition properly, large numbers of samples will have to be taken from many body sites at various time intervals after death. Furthermore, as explained in Chapter 4 (section 4.1), in order to make the data obtained statistically significant, a large number of different types of corpses (age, gender, race, cause of death etc.) will need to be investigated. Such studies are very time consuming and expensive and, basically, haven't yet been carried out – we're still only at an early stage in determining what microbes are present at all the different sites on a living human being, never mind a dead one. Those studies that have been published so far have involved a relatively limited number of corpses, a limited number of body sites and only a few time points following death. Furthermore, most of the studies that have been carried out have focussed only on bacteria and haven't determined which of the other major groups of microbes might be present. Consequently, we know very little about the communities of archaea, fungi, viruses and protozoa in corpses and how these vary at different stages. To summarise, our knowledge is very incomplete and we're still only at the beginning of research into this topic. However, it's a subject of great interest and a lot of research is going on to try and work out how the microbiotas of the various body sites change with time. This would, of course, be of great practical importance as it would give us another way of establishing the PMI (Box 5.4).

Box 5.4 Establishing the post mortem interval from microbiological data

As we'll see later in this chapter, the composition of the microbiotas of particular sites on a corpse change with time as decomposition progresses. The microbiotas of the mouth and gut, in particular, have been studied with a view to using changes in their composition as "microbial clocks" in order to estimate the PMI. However, such studies have been very few in number and have generally involved only small numbers of corpses. Another problem is that we know that several environmental factors affect the rate of corpse decomposition such as the climate, local weather conditions, the size and age of the deceased, the presence of clothing, type of burial etc. None of these factors have been studied with regard to their effects on the microbial communities present on a corpse. We are, therefore, a long way from being able to provide a PMI on the basis of a microbiological analysis of a corpse. Nevertheless, this is an exciting and promising area of research.

Because it's often difficult to get permission from relatives to carry out such studies on human corpses, many researchers have had to resort to studying the decomposition of other mammals - pigs and mice have proved useful in this respect. However, it's important to bear in mind that there are large differences between the microbiota of a particular body site in a human and in a pig and a mouse.

Unfortunately, at the time of writing (during the great COVID-19 plague of 2020-21) there have been very few published studies looking at the effects of gender, ethnicity or age on the microbiology of corpse decomposition. Neither have there been many studies published on the effects of climate, different soil types or type of burial on the corpse microbiota. As a minimum we need to know how the microbial communities of the various body sites vary at different time points after death but only a limited number of such studies have been published and these have involved only a very small number of corpses.

As was pointed out in Chapter 2, the microbial communities that inhabit the various regions of a live human differ enormously and it's difficult to describe the human microbiota as a whole, other than in general terms. We can only really gain an understanding of the human microbiota by looking at the microbiotas of individual regions, organ systems or sites. The same applies to understanding the decomposition of human corpses – it's best to look at what happens at individual regions of the body. What I'm going to do, therefore, is make a few generalisations about the corpse microbiota and then describe how the microbiota of particular body sites or organs change with time after death. Finally, I'll round this off by describing the role that environmental microbes play in the decomposition process. In Chapter 2 the point was made that the vast majority of the human microbiota live in the intestinal tract and that the skin and respiratory tract also support large microbial communities. Not surprisingly then, most studies of the corpse microbiota have focussed on the microbial (mainly bacterial) communities that inhabit these three sites.

5.6.1   A General Overview – The Broad-Brush Approach

Generally, the microbial communities at the various body sites don't alter dramatically during the first 24-48 hours after death. However, autolysis (as described previously) and changes in the physical characteristics of the corpse (e.g. decreased oxygen, pH alterations, temperature changes) then exert their effects resulting in very different environments. Such changes, together with the rapid degeneration of the antimicrobial defence systems in the corpse, mean that the microbiotas of the various sites now alter and these changes will be described later. After prolonged periods of decomposition, the microbiotas of different body sites generally become more similar to one another. Also, they become more similar to the microbial communities of the environment in which the corpse has been placed.

It's important to remember that, in addition to those body sites that are colonised by microbes (skin, gut, respiratory tract etc), there are many sites within the body of a human that don't have a microbiota. All of the internal organs (heart, liver etc), as well as internal tissues, are microbe-free during life, so what happens to them after death? Well, after autolysis has occurred, these organs and tissues are a "land of milk and honey" for microbial settlers. Also, the antimicrobial defence systems that protected them from the odd microbe that ventured there during life are no longer working and aren't there to fight off any microbial invaders. As explained previously, the thin mucosal layers that separated these sterile organs and tissues from the huge microbial communities of the gut, skin, respiratory tract etc., will have been damaged by autolysis and so are no longer effective barriers. Consequently, the previously-sterile internal organs and tissues now experience a microbial invasion. Which microbes colonise each of these organs will depend on a number of factors such as which of the microbe-rich regions is the closest and the nature of the environment each organ provides.

In Chapter 3 we saw that most of the microbes in a living human reside in the large intestine. The autolysis-induced damage to the thin mucosa that separates these gut microbes from the sterile internal tissues enables the migration (often termed "translocation") of these microbes. Although translocation hasn't been studied in human corpses, in mice it's been shown that intestinal bacteria invade sites such as cardiac blood, kidney, liver, mesenteric lymph nodes and spleen as early as 5 minutes after death. Within 5-48 hours, *Clostridium* species from the intestinal tract can be found at many sites within a human corpse. The species that are most frequently detected are *Cl. sordellii, Cl. difficile, Cl. bartlettii, Cl. bifermentans, Cl. limosum, Cl. haemolyticum, Cl. botulinum* and *C. novyi* (Figure 5.7). Collectively, these produce a wide range of macromolecule-degrading enzymes as well as toxins that can kill human cells resulting in the release of more macromolecules and nutrients (Table 5.3) .

The prevalence of *Clostridium* species throughout the human corpse has been called the "postmortem *Clostridium* effect" (PCE). Several properties of

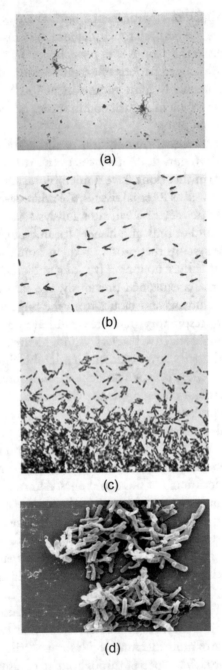

(a)

(b)

(c)

(d)

Figure 5.7 Images of some of the *Clostridium* species that are widely-dispersed throughout a human corpse soon after death
(a) *Clostridium novyi* stained to show the presence of flagella which are used by the bacterium (and most other *Clostridium* species) to move
Dr. William A. Clark, Centers for Disease Control and Prevention, USA

Table 5.3 Macromolecule-degrading enzymes and toxins produced by those *Clostridium* species that are widely-dispersed throughout a human corpse soon after death (+ = enzyme/toxin produced in that species; - = enzyme/toxin not produced or information not available)

Enzyme/toxin	*Cl. sordellii*	*Cl. difficile*	*Cl. bifermentans*	*Cl. haemolyticum*	*Cl. botulinum*	*Cl. novyi*
Lipase	+	+	+	+	+	+
Sialidase	+	-	-	-	-	-
Collagenase	+	+	-	-	+	+
Protease	+	+	+	+	+	+
Hyaluronidase	+	+	-	-	-	-
Fibrinolysin	+	-	+	-	-	-
Chondroitin sulphatase	-	+	-	-	-	-
Heparinase	-	+	-	-	-	-
Deoxyribo-nuclease	-	-	+	-	+	-
Xylanase	-	-	+	-	-	-
Haemolysin	+	+	+	-	+	+
Cytotoxin	+	+	-	+	+	+

Clostridium species account for their being able to spread throughout the corpse so rapidly and establish themselves at different body sites and these are summarised in Table 5.4.

The large intestine is an anaerobic environment that contains huge numbers of an enormous variety of anaerobic bacteria. Collectively these bacteria, as well as the fungi and protozoa that live there, produce a wide range of hydrolytic enzymes (see Section 3.1.8.4.) that can break down all of the macromolecules present in tissues. Many of the simple sugars, amino acids, fatty acids, and nucleotides produced by these enzymes then undergo fermentation (an energy-generating process that occurs under anaerobic conditions) resulting in the production of gases such as CO_2, hydrogen sulphide, methane and ammonia. These gases accumulate within the abdomen and inflate it – this is characteristic of the bloat stage of decomposition (see Section 4.3.2). The pressure build-up is sufficient to force out fluids ("purge fluids") from the nose

---

Figure 5.7 (continued) (b) Gram stain of *Clostridium botulinum* showing Gram-positive bacilli. Note the red-staining spores in many of the bacteria.
Dr. George Lombard, Centers for Disease Control and Prevention, USA
(c) Gram stain of *Clostridium difficile* showing Gram-positive bacilli – the pale regions within the cells are spores
Dr. Gilda Jones, Centers for Disease Control and Prevention, USA
(d). Scanning electron micrograph of *Clostridium difficile* (x3006) showing rod-shaped cells
Lois S. Wiggs, Centers for Disease Control and Prevention, USA

Table 5.4 Properties of *Clostridium* species that account for the postmortem clostridial effect

Property	Effect
They can grow and reproduce very rapidly	Some species can double in numbers every 7 minutes under optimum conditions therefore they can quickly outgrow other species of microbes.
They produce a wide range of hydrolases, particularly proteases	Enables them to break down a variety of macromolecules found in human tissues
They produce toxins that kill human cells (called "cytotoxins")	Releases nutrients for bacterial growth
Many are anaerobes, some are facultative anaerobes	They can grow under the anaerobic conditions found in the gut and that predominate in many other body regions in the early stages following death
They have flagella	This means they can move to other regions of the body
They have sensors that can detect nutrients	This enables them to seek out nutrients and to move to where they are plentiful
They can form spores	When they find themselves in conditions unsuitable for their growth they form spores which protect them. When the conditions improve the spores germinate and they can start growing and reproducing again.

and anus and eventually is sufficient to burst the skin and this marks the start of the active decay stage. Rupture of the abdomen will, of course, allow oxygen to penetrate into the corpse and this allows aerobes and facultative anaerobes to grow within the corpse. During this active decay stage, some microbes produce foul-smelling compounds such as cadaverine, putrescine and hydrogen sulphide and these are characteristic of this stage of the decomposition process.

5.6.2 What Happens to Specific Organs?

The various organs of the body decay at different rates depending mainly on whether or not they have a resident microbiota, the type of tissue they are composed of and how much blood they received during life. In general, decomposition occurs earliest in organs that contain less muscle and fibrous tissues and had an abundant blood supply. Muscular tissue has a high content of collagen and not many microbes have the ability to produce collagenase, the enzyme that's responsible for its breakdown. The general sequence of decay is shown in Table 5.5 and is illustrated for various organs in Figure 5.8.

Hair and nails are composed mainly of keratin which is a protein. Not many bacterial members of the human microbiota are able to produce the

Table 5.5 The sequence of decay of various organs and tissues in a human corpse listed in order of decreasing rapidity of decay

Organ/tissue	Organ system	Presence of a microbiota during life
Larynx and trachea	Respiratory	Yes
Stomach, intestine	Digestive	Yes
Spleen	Lymphatic	No
Mesentery	Digestive	No
Liver	Digestive	No
Pancreas	Digestive	No
Adrenal tissues	Endocrine	No
Pregnant uterus	Reproductive	Yes
Heart	Circulatory	No
Lungs	Respiratory	Yes
Kidneys	Urinary	No
Oesophagus, diaphragm	Respiratory	Yes
Blood vessels	Circulatory	No
Urinary bladder	Urinary	Yes
Bronchi	Respiratory	Yes
Prostate	Reproductive	No
Uterus of a female who's never been pregnant	Reproductive	Yes
Skin	Integumentary	Yes
Muscles, tendons	Muscular	No
Bones	Skeletal	No

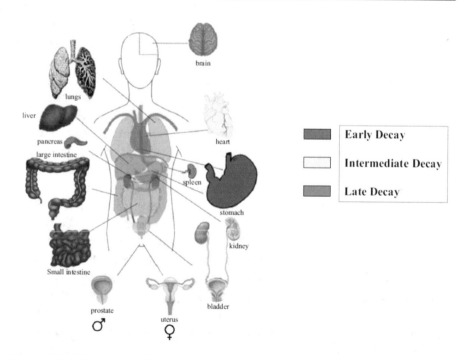

Figure 5.8 The order of decay of various organs in a human corpse
Image from: An interdisciplinary review of the thanatomicrobiome in human decomposition. Javan GT *et al. Forensic Science Medicine and Pathology* 2019 Mar;15(1):75-83. doi: https://doi.org/10.1007/s12024-018-0061-0

enzymes (keratinases) needed to hydrolyse keratin. This ability is more frequently found in fungal inhabitants of humans such as species from the genera *Aspergillus, Malassezia, Candida, Cladosporium* and *Fusarium*. Hair and nails, therefore, tend to persist for much longer and are frequently still present in the advanced and skeletal decay stages. Subsequent degradation is carried out by microbes from the environment. Many of the microbes present in soil can degrade keratin and these include bacteria (*Streptomyces, Bacillus, Actinomyces, Pseudomonas*) as well as fungi such as *Fusarium, Aspergillus, Penicillium, Acremonium* and *Geotrichum*.

5.6.2.1   What Happens in Your Mouth?

The mouth is one of those body sites that, during life, is colonised by many different types of microbe and the communities present have been described in Section 3.1.8.1. Death dramatically alters the environment of the mouth and this affects the microbiota which continues to change as the corpse decays. So, what causes the environment of the mouth to change after death? First of all, autolysis releases many additional nutrients for the microbial communities that live in the mouth. Particular species can benefit from these changes and so will be able to grow and reproduce more quickly while others won't be able to thrive and their numbers will decrease. Next, the saliva that usually bathes all of the surfaces in the mouth stops being produced and this is a valuable source of nutrients for many species and these will suffer. No more food enters the mouth after death and many species depend on the nutrients (particularly the sugars) that are present in this. While we're alive, saliva continually removes microbes and their waste products – these are swallowed and so end up in the stomach. This no longer happens, so microbes, as well as the acids and other waste compounds they produce, will accumulate in the mouth and alter its environment. Then, of course, the mouth will be invaded by microbes from the respiratory tract, gut and the external environment and so will have to compete with these microbes for nutrients. The microbiota of the mouth therefore will change and subsequent community succession (described above) occurs as the corpse passes through the various stages of decay (Figure 5.9).

During the fresh stage (Figure 5.9a), the microbiota of the mouth is similar to that found in living humans which has been described in detail in Section 3.1.8.1. However, once the bloat stage is reached (Figure 5.9b) the microbiota contains, in addition to oral bacteria, large proportions of bacteria from the colon such as *Enterococcus* and *Vagococcus* (Box 5.5) and from the environment (*Acinetobacter*) (Box 5.6). Bacteria associated with insects, such as *Ignatzschineria*, also start to appear at this stage – these will have been deposited on the corpse by visiting insects. The active decay stage (Figure 5.9c) is

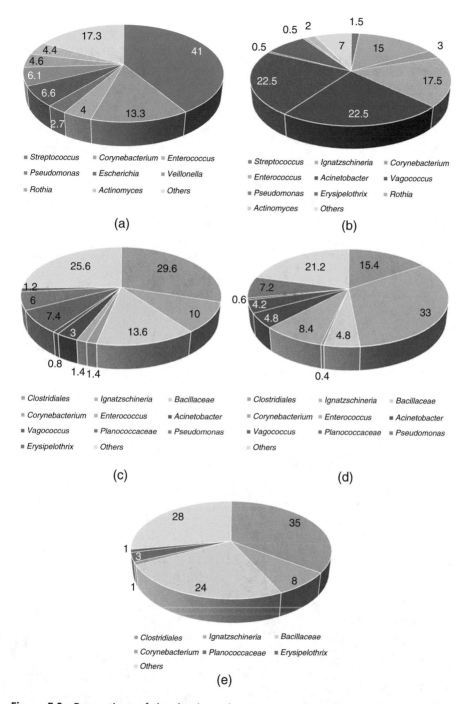

Figure 5.9 Proportions of the dominant bacteria present in the mouth at different stages of decomposition: (a) fresh, (b) bloat, (c). Active decay, (d) advanced decay (e). Skeletonization. The figures are the relative proportions (%) of the various bacteria found in several individuals

Dynamics of the oral microbiota as a tool to estimate time since death. J. Adserias-Garriga et al. *Molecular Oral Microbiology* 2017;32:511-516

Box 5.5 Some important examples of bacteria that live in the human gut

Enterococcus species are facultatively anaerobic Gram-positive cocci that usually occur in pairs or chains (Figure a). They can grow over a wide temperature range (5-65°C) and can grow under a variety of conditions which makes them very versatile. The ability to grow at low temperatures means they can grow and reproduce at temperatures found in a human corpse. They are found in the environment (soil, sand, water, on vegetation) and are usually present in the intestinal tract of humans and other animals. They can break down proteins, hyaluronic acid, lipids, polysaccharides, nucleic acids and can kill human cells.

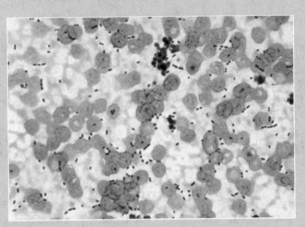

Figure (a) Gram stain of an *Enterococcus* species showing Gram-positive cocci that are mainly in pairs and short chains. Much larger, red-staining human cells can also be seen
Dr. Mike Miller, Centers for Disease Control and Prevention, USA

Vagococcus species are Gram-positive facultatively anaerobic motile cocci. They are found in the intestinal tract of humans and other animals as well as in soil and water. They can grow at temperatures as low as 10°C which enables them to grow and reproduce at temperatures found in a human corpse. They can break down polysaccharides, peptides and lipids.

Clostridiales is an order within the Firmicutes phylum and consists of 243 genera in 16 families. They are Gram-positive anaerobic bacilli. The main genera found in the human gut are *Clostridium, Sarcina, Eubacterium, Peptostreptococcus, Peptococcus, Filifactor* and *Ruminococcus*. They are generally very effective at breaking down proteins.

Faecalibacterium is a genus of Gram-negative, anaerobic, non-motile bacilli. It comprises a high proportion of the gut microbiota of most adults. It can hydrolyse a wide range of polysaccharides, proteins and nucleic acids.

Phascolarctobacterium is a genus of Gram-negative, anaerobic, non-motile bacilli that is frequently present in the human gut.

Lactobacillus is a genus of Gram-positive, anaerobic, non-motile bacilli. They are invariably present in the human gut where they break down proteins, lipids, DNA and polysaccharides.

Box 5.5 (continued)

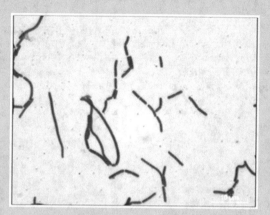

Figure (b) Gram stain of a *Lactobacillus* **species (x1000) showing Gram-positive bacilli**
Nishida S et al. *Lactobacillus paraplantarum* 11-1 Isolated from rice bran pickles activated innate immunity and improved survival in a silkworm bacterial infection model. *Frontiers in Microbiology* 2017; 8:436. doi: https://doi.org/10.3389/fmicb.2017.00436. This is an open-access article distributed under the terms of the Creative Commons Attribution License (CC BY).

Blautia is a genus of Gram-positive, anaerobic, non-motile bacilli that is frequently present in the human gut. It can break down proteins, lipids and carbohydrates.

Bacteroides is a genus of Gram-negative, anaerobic, non-motile bacilli that is invariably present in the human gut where it usually comprises a high proportion of the microbiota. They can break down a wide range of polysaccharides and proteins as well as mucins and glycoproteins. Also of importance is their ability to break down extracellular tissue macromolecules such as heparan and hyaluronan.

Figure (c) Gram stain of a *Bacteroides* **species showing Gram-negative bacilli**
Don Stalons, Centers for Disease Control and Prevention, USA

Box 5.6 Some important types of bacteria that are primarily found in the external environment

Bacillaceae is a family of Gram-positive bacilli that are usually motile and usually produce spores. They are widely distributed in the environment and are found in soil, freshwater, seawater and hotsprings. The family consists of 457 species in 31 genera. The largest genus, *Bacillus*, consists of 226 species. They are very effective at breaking down polysaccharides, proteins, lipids and nucleic acids.

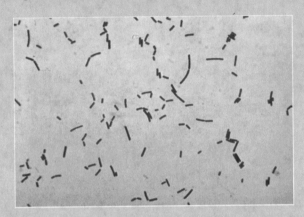

Figure (a) Gram stain of a *Bacillus* species showing Gram-positive bacilli Dr. William A. Clark, Centers for Disease Control and Prevention, USA

Planococcaceae is a family of Gram-positive cocci that consists of 14 genera. They are aerobic and some genera are motile and some can form spores. They are widely distributed in the environment and are found in soil, freshwater and seawater. They produce proteases and lipases.

Acinetobacter species are Gram-negative bacilli (Figure b) that can only grow in the presence of oxygen. They are widely distributed in the environment and are found in soil, water, sewage, animals (birds and fish) and on human skin. They can break down lipids, polysaccharides and peptides and can kill red blood cells and other human cells thereby releasing more nutrients for themselves.

Box 5.6 (continued)

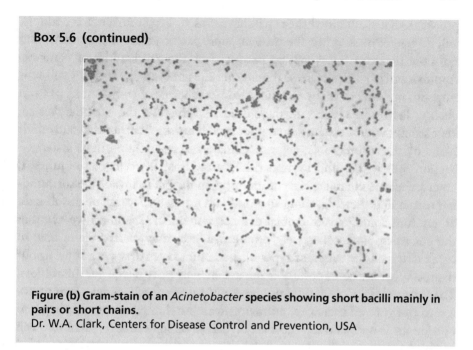

Figure (b) Gram-stain of an *Acinetobacter* species showing short bacilli mainly in pairs or short chains.
Dr. W.A. Clark, Centers for Disease Control and Prevention, USA

marked by high proportions of bacteria from insects (*Ignatzschineria*) as well as from the colon (*Clostridiales*) and from the soil (*Bacillaceae, Planococcaceae*). During the advanced stage of decay, bacteria from the soil and insects predominate (Figure 5.9d). Soil bacteria come to dominate the mouth during the skeletonization stage (Figure 5.9e).

5.6.2.2 What Happens in Your Gut?

Like the oral cavity, the gut has a large microbiota while we're alive and, as mentioned previously (Chapter 3), it contains most of the microbes that accompany us during our life. Death results in huge changes to the environment within the gut and so we can expect the microbial communities living there to alter significantly as well. First of all, once we've died no more food enters the gut and this is the main source of nutrients for the microbes that live there. Autolysis will provide nutrients for a short period after death, as will the digested food that was present before death. The gut as a whole is a series of distinct regions – the stomach, small intestines, caecum and large intestines (Chapter 3). Each of these regions has a distinct environment but, following death, these distinctions disappear because the physiological processes that maintained them will have ceased. The stomach, for example, will no longer be so acidic (because no more gastric juice is being made) and no more digestive juices will be produced by the gall bladder or pancreas. Approximately 7 litres of these fluids enter the gut each day and this will now stop. So, not only does the gut suffer from a shortage of water but its pH will change due to the absence of these fluids. During life, the gut microbes are kept strictly within the intestinal tract by the mucosal barrier that surrounds it and by the antimicrobial defences (innate and acquired immune systems – Section 3.2) within and beyond this. Autolysis of the cells of the mucosa, however, allows the microbial communities in the gut to invade the surrounding tissues and there's nothing to stop them doing so because of the absence of any functioning antimicrobial defence system. In Chapter 3 we saw that the gut microbiota collectively produces a vast array of hydrolases able to break down all the different types of macromolecules they encounter in our tissues. What is also important is the extreme diversity of the gut microbiota. There are at least 17,000 different types of bacteria, fungi, archaea and protozoa in the gut and therefore, no matter what kind of environment they find outside the intestinal tract, there are certain to be some among them that will be able to survive and grow there. It's this versatility that enables members of the gut microbiota to spread throughout the corpse and colonise all regions.

Very few studies of the gut microbiota of corpses have been carried out and Figure 5.10 summarises the results of one investigation that involved taking samples at various times from three corpses left out in the open air.

The proportions of Firmicutes (mainly *Faecalibacterium, Phascolarctobacterium, Lactobacillus* and *Blautia*) and Bacteroidetes (mainly *Bacteroides* and *Parabacteroides*), which are all characteristic of the gut microbiota of live humans (Box 5.5) remain fairly constant until the middle of the bloat phase (days 4-7 in the three corpses). After this, their proportions declined throughout the remaining decay stages. In contrast, the proportions of other Firmicutes (*Clostridium, Peptostreptococcus, Anaerosphaera*) and Proteobacteria (*Providencia, Acinetobacter, Wohlfahrtiimonas* and *Ignatzschineria)* increased after the mid-bloat stage. While species belonging to the first four of these genera would have been present in the human gut during life, species from the genera *Wohlfahrtiimonas* and *Ignatzschineria* would have been introduced by insects. *Acinetobacter* species are generally found in soil and water and so represent the invasion of the corpse by environmental microbes. The proportion of another genus of bacteria found in the human gut during life, *Bifidobacterium*, generally remained steady throughout the five stages of decay. After 16 days, which is in the skeletonization stage of decomposition, the Firmicutes (mainly *Clostridium, Peptostreptococcus* and *Anaerosphaera*) became dominant and comprised between 73 and 98% of the microbiota.

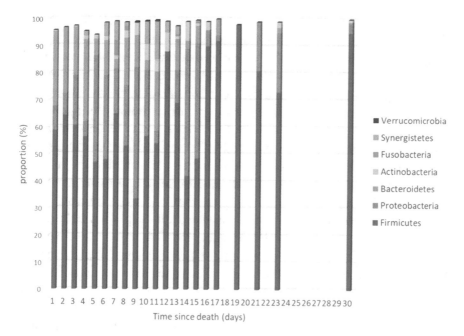

Figure 5.10 Composition of the gut microbiota at various times since death. The study involved the sampling of 3 corpses and the figures show the proportions (%) of the main phyla of bacteria detected

The main genera detected in each of the bacterial phyla that dominated the microbiota were as follows:

Firmicutes: *Faecalibacterium, Phascolarctobacterium, Lactobacillus, Blautia, Clostridium, Peptostreptococcus, Anaerosphaera*

Bacteroidetes: *Bacteroides, Parabacteroides*

Proteobacteria: *Wohlfahrtiimonas, Ignatzschineria, Acinetobacter, Providencia*

5.6.2.3 What Happens to Your Skin?

The complex microbiota of the skin has been described in Section 3.2.4. During life, the main nutrients for the microbes living on the skin come from the sebum and sweat produced by the sebaceous and sweat glands respectively, the fats and other macromolecules produced by skin cells and the waste products of other skin microbes. Following death, no more sebum or sweat will be produced but autolysis of skin cells acts as an alternative source of nutrients for skin microbes. The skin, being our outer boundary, is in contact with the external environment more than any other region of our bodies. During life, the body's antimicrobial defence systems are able to selectively exclude microbes from the external environment and this exclusion is helped by the antimicrobial compounds produced by the resident microbes of the skin (see Box 5.2). Following death, however, the body's defence systems no longer operate and so the task of repelling environmental microbes is left to our microbial symbionts. While we are alive our skin is unlikely to come into contact with huge numbers of environmental microbes – unless we choose to engage in mud-wrestling or other activities that result in us being covered in earth. Most of our skin will encounter only the occasional microbe from the surrounding air – this contains only about 10^6 microbes per m^3. In contrast, soil contains in the region of 10^{15} microbes per m^3 (although this figure varies enormously depending on soil type, climate etc.) i.e. approximately 1,000,000,000 times more. So, our symbionts are faced with enormous numbers of environmental microbes with an interest in accessing the nutrients present in a human corpse. Not only do environmental microbes greatly outnumber our skin microbes but they have another important advantage – they are already adapted to the conditions in which the corpse has been placed. Take temperature, for example. The microbes that live on our skin are used to a temperature of around 37°C, thanks to their generous host. Death, of course, puts paid to that and the temperature of the skin will have fallen (in most climatic regions) well below that. This means that the growth of skin microbes will slow down and they'll be reproducing at a much lower rate. In contrast, environmental microbes are adapted to the prevailing temperature (and other conditions found there, such as humidity) because this is their normal environment and so will be able to grow and reproduce at their usual rates. We can, therefore, expect the skin to be quickly colonised, and decomposed, by microbes from the external environment.

Autolysis of the epithelial cells of the skin severely disrupts its ability to exclude microbes (from the skin and the environment) from the underlying tissues and so we can expect invasion of these previously-sterile sites.

We really know very little about how the microbial communities present on skin change during corpse decomposition. However, the results of one study carried out on two corpses decomposing in the open air are summarised in Figure 5.11. Samples were taken from the skin above the left and right biceps over a 24 day period that included all five major decomposition stages. During the fresh and bloat stages (up to 4 days) soil bacteria (*Acinetobacter* and *Pseudomonas*) were the dominant bacteria present, although skin bacteria (*Staphylococcus* and *Corynebacterium*) were also found. During the active decay stage (4-8 days) the proportions of *Ignatzschineria* (from insect larvae) and *Clostridium* (from the gut) were high. The advanced (dry) stage (10-20 days) saw decreasing proportions of *Ignatzschineria* due to the decline in maggot activity (see Chapter 6) and increasing proportions of bacteria from the gut (*Clostridium*). At skeletonization (day 24), the microbial communities were dominated by gut bacteria (*Clostridium*) and soil bacteria (*Acinetobacter*).

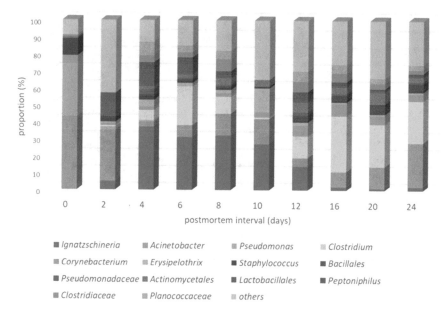

Figure 5.11 Composition of the skin microbiota at various times since death. The study involved the sampling of the skin from the left and right arm (over the biceps) of two corpses and the figures show the proportions (%) of the main groups of bacteria detected

5.6.2.4 What Happens in Your Brain?

The brain is a sterile organ during life. Although brain cells die within 3-7 minutes after death, the DNA inside them can remain largely intact for as long as 3 weeks after death. Although it's situated a long way from the intestinal tract where most of the tissue-destroying microbes live, it's close to the mouth and nasal cavity which have large microbial communities.

Few studies have looked at the microbiota of the brain while it decays. Figure 5.12 shows the results of one study that looked at the microbiota of the brain of males and females at different times during the early stages of decay.

Such investigations involve a considerable amount of work and this limits the number of corpses that can be studied. As only 4 males and 5 females were sampled it would be unwise to generalise too much from these results. The data show that while anaerobic bacteria (an unknown member of the *Clostridiales* and *Clostridium* species) are the dominant bacteria in this early stage of decay in both males and females, there are gender-associated differences. The most striking is that *Pseudomonas* species comprise a substantial proportion of the microbiota in females, but not in males. In contrast, the proportion of *Streptococcus* species is high in males but not in females. *Enterobacteriaceae* comprise a substantial proportion of the microbiota in the

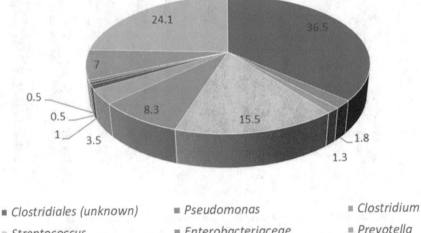

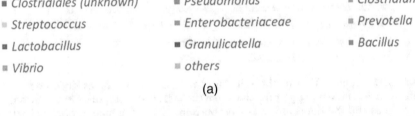

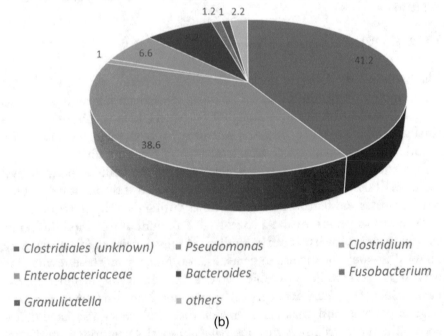

Figure 5.12 The proportions of the main types of bacteria detected in the brains of four males (a) and five females (b) during the early stages of decomposition. The PMI of all corpses was less than 2.5 days

Human thanatomicrobiome succession and time since death. *Scientific Reports* 2016; 6:29598 | DOI: https://doi.org/10.1038/srep29598

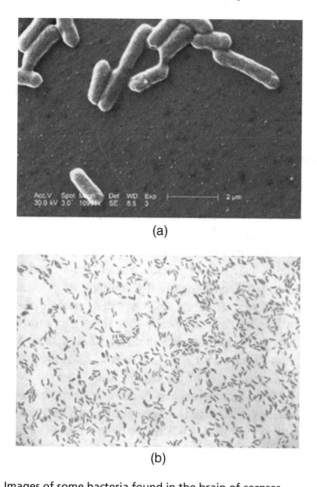

(a)

(b)

Figure 5.13 Images of some bacteria found in the brain of corpses
(a) *Escherichia coli*, a typical member of the *Enterobacteriaceae* as seen through an electron microscope (X10,961). It is a rod-shaped bacterium
National Escherichia, Shigella, Vibrio Reference Unit at Centers for Disease Control and Prevention, USA
(b) Gram stain of a *Pseudomonas* species showing Gram-negative bacilli
Dr. W.A. Clark, Centers for Disease Control and Prevention, USA

brains of both genders. *Enterobacteriaceae* are a large family of Gram-negative bacteria that are plentiful in the human gut and so this is their most likely origin. They are facultative anaerobes, most are motile and collectively they produce a wide range of hydrolases. *Pseudomonas* species are aerobic bacteria (Figure 5.13) that are present in the gut but are also found in the environment. They are motile and produce a wide range of macromolecule-degrading enzymes including proteases, lipases, deoxyribonucleases and elastase.

Streptococcus species are among the dominant bacteria found in the mouth, stomach and respiratory tract. They are facultative anaerobes and produce enzymes that can hydrolyse proteins, polysaccharides, glycoproteins, hyaluronan, mucins, chondroitin sulphate and DNA.

After 10 days, i.e. during the skeletonization stage, anaerobic bacteria from the gut comprised around 84% of the microbiota.

5.6.2.5 What Goes on in Your Heart (With Thanks to the Beatles)?

The heart, like the brain, is free of microbes during life. Although it's close to a major source of microbes (the respiratory tract), it's quite resistant to decomposition. Cardiac tissue can remain intact for as long as 4 days after death and the coronary arteries may be recognisable several weeks later. Such resistance is probably due, in part, to the heart's high content of muscular and fibrous tissue. In those who suffered from atheroma (hardened arteries) during life, the calcification of their arteries means that these can survive even longer because they are more difficult for microbes and insects to breakdown. Anaerobic bacteria (*Clostridiales* and *Clostridium*) from the gut dominate the microbiota of the early stages of the decomposing heart in both males and females (Figure 5.14). Interestingly, as in the case of the brain, *Pseudomonas* species appear to be plentiful in the hearts of females (Figure 5.14b) during the early stages of decomposition – this has been reported in at least two different studies.

During the skeletonization stage, after about 10 days, anaerobic bacteria from the gut comprised around 98% of the microbiota.

5.6.2.6 What Happens to Your Liver?

The liver (approximately 1.4 Kg) is the heaviest internal organ and the largest gland in the body – during life it's sterile. It decays very rapidly after death and this is due to a number of factors. Firstly, the liver is very close to the huge microbial communities that inhabit the large intestine. Secondly, autolytic breakdown of the liver tissues is supplemented by the cell-damaging effects of acidic fluid from the stomach (which is very close) and enzymes from the pancreas (also very close). This means that the intestinal microbes, when they arrive, are provided with plenty of sugars, amino acids and other breakdown products of the macromolecules that made up the liver. It's one of the most microbially-diverse organs during decomposition.

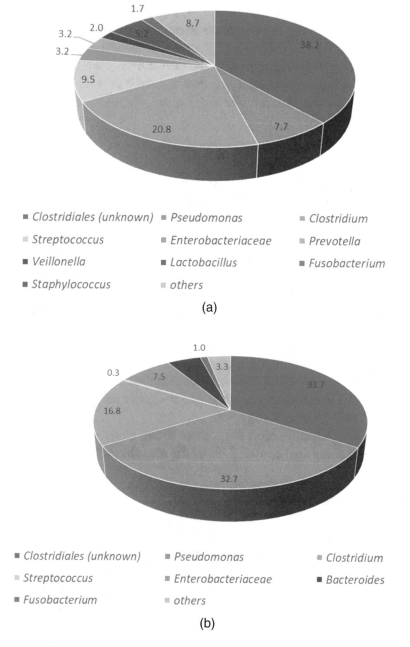

Figure 5.14 The proportions of the main types of bacteria detected in the hearts of four males (a) and six females (b) during the early stages of decomposition. The PMI of all corpses was less than 2.5 days
Human thanatomicrobiome succession and time since death. *Scientific Reports* 2016; 6:29598 | DOI: https://doi.org/10.1038/srep29598

As with the other internal organs, anaerobic species (*Clostridiales* and *Clostridium*) are among the dominant bacteria in the decomposing livers of both males and females (Figure 5.15). Again, *Pseudomonas* species comprise a high proportion of the microbiota in the case of livers from females.

During the skeletonization stage, after about 10 days, anaerobic bacteria from the gut comprised around 98% of the microbiota.

5.6.2.7   What Happens to Your Spleen?

The spleen (about 250 g) is the largest ductless gland of the body and is sterile during life. It filters and stores blood and produces important components of our immune defence system. It's known as a ductless gland because, unlike glands such as the ones that produce saliva and sweat, it releases the substances it produces directly into the bloodstream rather than via ducts. It has a plentiful blood supply and so readily undergoes decomposition. It rapidly become discoloured after death due to the large quantities of sulphaemoglobin produced by the reaction of haemoglobin with microbially-produced hydrogen sulphide. It lies very close to the large intestine which contributes to it being one of the most microbially-diverse organs during decomposition.

As with the other internal organs, *Pseudomonas* species comprise a high proportion of the microbial communities of the spleens of females during the early stages of decomposition, while gut anaerobes are predominant in the spleens of males (Figure 5.16)

During the skeletonization stage, after about 10 days, anaerobic bacteria from the gut comprised around 97% of the microbiota.

5.6.2.8   What Happens to the Uterus?

Only recently has the uterus been shown to have a resident microbiota although the numbers of microbes present are relatively small. This, combined with the fact that it has a high content of muscle and fibre, means that it's less susceptible to decomposition. No studies have been published regarding the microbial communities present in the uterus of a corpse.

5.6.2.9   What Happens to the Prostate?

Like the uterus, the prostate is slow to decompose and is one of the longest-surviving organs after death. The reasons for this aren't clear but its high content of difficult-to-degrade muscle and fibre are contributing factors – it's

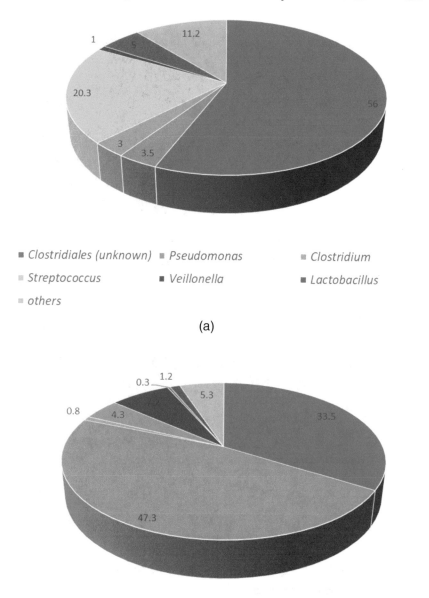

Figure 5.15 The proportions of the main types of bacteria detected in the livers of four males (a) and six females (b) during the early stages of decomposition. The PMI of all corpses was less than 2.5 days

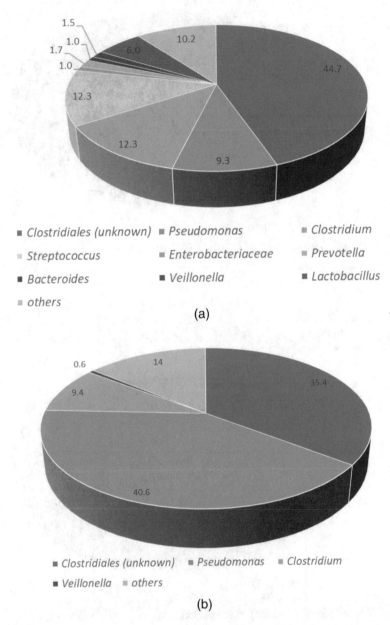

Figure 5.16 The proportions of the main types of bacteria detected in the spleens of six males (a) and five females (b) during the early stages of decomposition. The PMI of all corpses was less than 2.5 days

composed mainly of muscle and has a protective sheath of collagen around it. Recently another possible contributory factor has been revealed – the ability of prostate cells to inhibit autolysis. A study of gene expression in prostate cells after death has shown that genes which prevent autolysis are functioning for at least 5 days after death. Consequently the cells that make up the prostate, unlike those of most other organs, don't immediately die and release nutrients for invading microbes and this means that this organ is far less susceptible to microbial decomposition.

5.6.2.10 What Happens to Them Bones, Them Bones, Them Dry Bones (With Thanks to the Johnson Brothers)?

In Section 2.1.2 we saw that the main constituents of bone are a mineral (hydroxyapatite), a protein (collagen) and small quantities of other macromolecules. Its decomposition is known as "diagenesis". Hydroxyapatite can't be broken down by enzymes but can be dissolved by the acids that are produced by microbes. However, collagen can be degraded to amino acids by collagenase enzymes produced by microbes. Either of these processes will weaken the integrity of bone and contribute to its eventual decomposition. While acid is produced by a huge range of bacterial and fungal members of the human microbiota, collagenase production is more limited. Examples of bacteria from the human microbiota that can produce collagenases include species from the genera *Alcaligenes, Bacillus, Pseudomonas* and *Clostridium* (see also Table 3.5). Fungal collagenase-producers found in the human microbiota include species from the genera *Mucor, Aspergillus, Cladosporium* and *Penicillium*.

Decomposition of bones takes considerably longer than that of other body organs and there is little evidence of any changes during the first few weeks following death. The reduction of a corpse to a skeleton generally takes several months and the microbes present in the bones of corpses at three different stages of skeletonization have been investigated and the results are shown in Figure 5.17. During the partially-skeletonised stage when larvae were still active and the corpse still retained liquid, the microbiota was dominated by *Pseudomonadaceae* and Firmicutes (*Clostridiaceae* and *Tissierellaceae*). The former are likely to have come from the soil while the Firmicutes are gut bacteria. Later, during the skeletonised stage when the corpse was still moist but no larvae were active, *Pseudomonadaceae* and *Clostridiaceae* remained the dominant bacteria. However, the microbiota was far more complex and contained small proportions of a wide variety of other bacterial groups (Figure 5.17b). By the time the skeleton had become completely dry (Figure 5.17c), the proportion of gut bacteria was considerably reduced. The microbial communities were far more varied and contained high proportions of a variety of environmental bacteria such as *Pseudomonadaceae, Caulobacteraceae, Sphingomonadaceae, Xanthomonadaceae, Streptomycetaceae* and *Bacillaceae*.

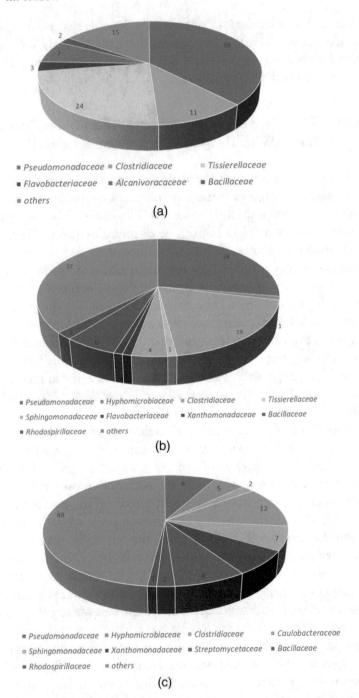

Figure 5.17 Microbiota of rib bones from 12 corpses left in the open air. Samples were taken at different stages of skeletonization (a) partially-skeletonised (PMI = 1-7 months), (b) skeletonised (PMI = 9-12 months), (c) dry remains (PMI = 18-48 months)

Interestingly, the bone-degrading microbes form tunnels in the bone (Figure 5.18).

5.6.2.11   What Happens to Your Blood?

The microbiota of blood in a corpse can only be studied during the earlier stages of decay because it eventually dries up, or becomes very difficult to sample, during the later stages. Table 5.6 summarises the main organisms found during these earlier stages and indicates their likely origin.

It can be seen from this table that a variety of bacteria can be detected in blood soon after death. During this fresh phase, it's difficult to specify exactly where these bacteria have come from because all of the species detected are found in the mouth, respiratory tract and the gut during life. However, during the bloat and active decay stages the situation is clearer because most of the species present are those that are found mainly in the gut. The *Cl. botulinum* detected during the active decay stage is an exception because this organism is not usually present in the gut but comes from the soil.

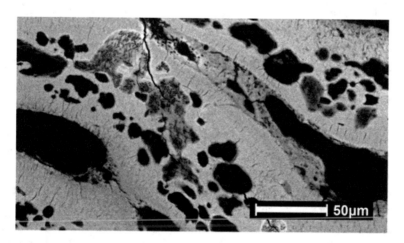

Figure 5.18   Scanning electron micrograph of a fragment of bone after one year. The tunnelling due to microbial decomposition can be seen (dark regions) and these extend to a depth of 200 μm

Eriksen AMH. *et al.* (2020) Bone biodeterioration—The effect of marine and terrestrial depositional environments on early diagenesis and bone bacterial community. *PLoS ONE* 15(10): e0240512. https://doi.org/10.1371/journal.pone.0240512

Table 5.6 Microbes detected in blood during the earlier stages of decomposition. The table lists only the dominant bacteria detected, many other species are also present in lower numbers

Decay stage	Main organisms present	Likely origins of the microbes found
Fresh	*Lactobacillus, Veillonella, Prevotella, Streptococcus, Gemella*	All of these are present in the gut as well as the mouth and respiratory tract of live individuals
Bloat	*Clostridium sordellii, Clostridium difficile, Clostridium bartlettii, Clostridium bifermentans, Clostridium limosum,*	All of these are present in the gut of live individuals
Active decay	*Clostridium haemolyticum, Clostridium novyi, Clostridium botulinum, Escherichia coli, Escherichia albertii*	All of these are present in the gut of live individuals except *Cl. botulinum* which is found in the soil

5.6.3 Come on in, It's Now Open House – The Invasion of the Body-Eaters

From the above descriptions of corpse decomposition, it can be seen that while our microbial symbionts are very active in the early stages, they are later helped by bacteria from insects (more about these in Chapter 6) and from the environment. Microbes from the environment become increasingly important as decomposition progresses and eventually they come to dominate the microbial communities during the later stages. While some of these microbes will have arrived from the air, the vast majority will have entered the corpse from the surrounding soil because, as pointed out previously, the soil has a huge population of microbes. Alternatively, if the body has been left in an aquatic environment (freshwater or seawater) then microbes present in that water body will have access to the corpse. As most of us will be buried in soil, further discussion will be confined to that situation.

We know that the microbial communities inhabiting the various body sites of humans are very similar regardless of where on planet Earth a human spends their life. While there are differences between males and females and between the young and the elderly, all humans generally have similar microbial communities. Therefore, we have a good idea of what microbes will be present at the time of death and so which microbes will start the decomposition of our corpse. However, the situation is different when we come to consider soil. It's very difficult to generalise about the microbial communities, as well as the other types of organisms (insects, worms etc.), that are likely to be present in the soil because these vary hugely depending on the type of soil in which the corpse is buried and the prevailing climate.

Soil is defined as the material on the surface of the Earth in which plants can grow. Basically, it's made up of various proportions of clay, silt and sand particles, the relative proportions of which give rise to its texture. Whatever the type of soil, it consists mainly of particles and pores with only a small proportion of organic material (Figure 5.19)

Of the organic material only approximately 5% consists of living organisms. Nevertheless, the organisms that live in soil are extremely important in maintaining its structure as well as its physical, chemical and biological properties. These organisms are usually classified into three main groups on the basis of their size (Table 5.7). The microbiota are the largest group of the three, both in terms of numbers and total weight, and about half of the total microbial population lives in the top 10 cm layer of the soil.

In this chapter we're concerned only with the microbiota of the soil, the insects will be discussed in Chapter 6 and other creatures in Chapter 7. Just as the human body contains a huge variety of habitats, so does the soil. Examples of the various regions in soil that have different environmental conditions include the surface of a particle, a water-filled pore, an air-filled pore and the surface of the root of a plant (known as the rhizosphere). Each of these will provide a habitat that is suitable for only certain types of microbes. Furthermore, a particle in the upper layers of the soil provides a different environment to one in the deeper layers because of differences in oxygen content, water content, temperature etc. Even when two soil particles are at the same depth in the soil they can still provide different environments if their chemical compositions are different. Consequently, as in the human body, there a wide variety of microbial communities present in the soil. This makes it very difficult to talk about the "typical" microbial content of a soil – there's really no such thing. However, in general, bacteria and fungi are usually the dominant types of microbes in soil and their total weight is about a thousand times greater than that of the viruses, archaea, algae and protozoa that are also present. Examples of bacteria and fungi that are frequently present in many soils and their main characteristics are given in Table 5.8. From this, it can be seen that all of the main groups of microbes found in soil are capable of breaking down a wide range of macromolecules present in the human corpse. The appearance of some of these microbes is shown in Figures 5.20 and 5.21.

As can be seen from Table 5.8, these environmental bacteria and fungi produce a wide range of enzymes that can break down many of the macromolecules present in human tissues. All of the bacteria and fungi listed can also produce keratinases which means that they can eventually break down the hair, nails and skin of the corpse, all of which have a high keratin content.

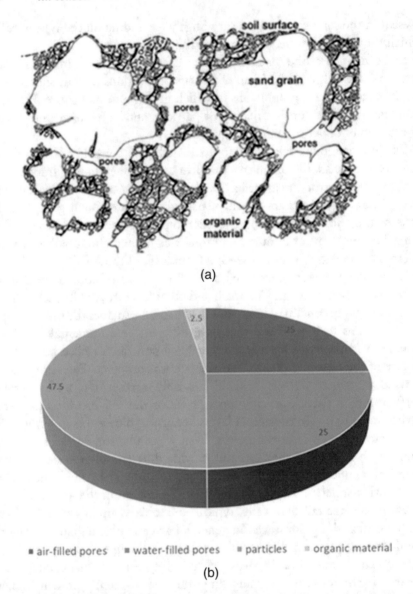

(a)

(b)

■ air-filled pores   ■ water-filled pores   ■ particles   ■ organic material

Figure 5.19 Soil structure and composition. In (a) the structure of soil can be seen to consist of particles and pores. (b) shows the general composition of soil. Living organisms comprise about 5% of the organic material present in soil
(a) Courtesy of State of Victoria (Agriculture Victoria), Australia. Creative Commons Attribution 4.0 licence

These environmental microbes have a great advantage over species from the human microbiota in that they are adapted to the prevailing climatic conditions in which the corpse has been placed. For example, they are all capable of growing at the lower temperatures found in the external environment, unlike human symbionts which grow best at 37°C.

Table 5.7 The main groups of organisms that live in soil

Group	Size	Main examples
Microbiota	Less than 0.2 mm	viruses, archaea, bacteria, fungi, algae and protozoa.
Mesobiota	0.2 to 10 mm	nematodes (round worms), enchytraeids (white worms), collembola (springtails), mites, rotifers and small insects
Macrobiota	Larger than 10 mm	earthworms, mollusks and large insects.

Table 5.8 Brief summary of the characteristics of major genera of bacteria and fungi that are frequently present in soil

Organism	Appearance	Relationship to oxygen	Macromolecule-degrading enzymes produced
Bacillus	Gram-positive bacilli, motile, form spores	Aerobes or facultative anaerobes	Proteases, glycosidases, lipases, nucleases, hyaluronidases
Pseudomonas	Gram-negative bacilli, motile	Aerobes	Proteases, lipases, nucleases, sialidases, hyaluronidases
Acinetobacter	Gram-negative bacilli, motile	Aerobes	Proteases, lipases
Actinomyces	Gram-positive bacilli, form spores	Facultative anaerobes	Proteases, glycosidases, lipases, sialidases
Arthrobacter	Gram-positive bacilli	Aerobes	Proteases, lipases, sialidases
Aspergillus	A fungus - forms a mycelium and spores	Aerobes	Proteases, glycosidases, lipases, nucleases, sialidases
Fusarium	A fungus - forms a mycelium and spores	Facultative anaerobes	Proteases, glycosidases, lipases, nucleases
Geotrichum	A fungus - forms a mycelium and spores	Aerobes	Proteases, lipases, glycosidases
Mucor	A fungus - forms a mycelium and spores	Facultative anaerobes	Proteases, lipases
Penicillium	A fungus - forms a mycelium and spores	Facultative anaerobes	Proteases, glycosidases, lipases, nucleases

So, when do these environmental microbes start to invade a human corpse? This is a very difficult question to answer because it will depend so much on the type of corpse (size, weight etc), how the corpse is buried, whether or not the corpse is damaged (either before or after death), whether or not it is clothed, the type of soil, the prevailing climate etc. Because of all these variables, it's difficult to generalise. However, some environmental microbes can be detected in some regions of the corpse as early as during the fresh stage of

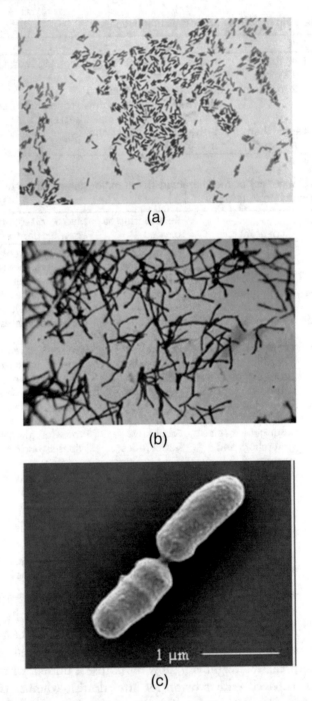

Figure 5.20 Microscopic appearance of some of the bacterial genera that are frequently present in soils

decomposition. As decomposition progresses, environmental microbes play an increasingly important role in the process until eventually they become the dominant microbes during the later stages. Those parts of the body (hair, nails and bones) that our microbial symbionts find difficult to break down will ultimately undergo decomposition due to the activities of environmental microbes. Just how important environmental microbes are in the decomposition of mammals has been shown in a recent study using mice. A group of dead mice were placed on ordinary soil and another group were placed on similar soil that had been sterilised. The mice placed on the ordinary, microbe-containing soil decomposed at a rate 2-3 times faster than those that were on the microbe-free soil.

Now that we have a good idea of how microscopic organisms are contributing to the recycling of our body, it's time to take a look at the role played by larger creatures in this process. The next chapter, therefore, will be devoted to insects.

5.7   Want to Know More?

Human Postmortem Microbiome Project (HPMP).
The HPMP endeavours to cultivate data that represents an extensive resource cataloguing the abundance and variety of microorganisms involved in humans (and/or human surrogates) decomposition.
https://hpmmdatabase.wixsite.com/hpmmdatabase
https://engagedscholar.msu.edu/magazine/volume14/learning-from-the-dead.aspx

---

Figure 5.20 (continued) (a). Gram stain of a *Bacillus* species showing Gram-positive bacilli
Dr. W.A. Clark, Centers for Disease Control and Prevention, USA
(b). Gram stain of an *Actinomyces* species (X1200) showing Gram-positive irregularly-shaped bacilli
Dr. Lucille K. Georg, Centers for Disease Control and Prevention, USA
(c). Scanning electron micrograph of an *Arthrobacter* species (x30,000) showing two short bacilli
Erythema caused by a localised skin infection with *Arthrobacter mysorens*
Imirzalioglu *et al. BMC Infectious Diseases* 2010; 10: 352. 2010 Dec 15. doi: https://doi.org/10.1186/1471-2334-10-352
This is an Open Access article distributed under the terms of the CreativeCommons Attribution License (<url>http://creativecommons.org/licenses/by/2.0</url>), which permits unrestricted use, distribution, and reproduction in any medium, provided the original work is properly cited

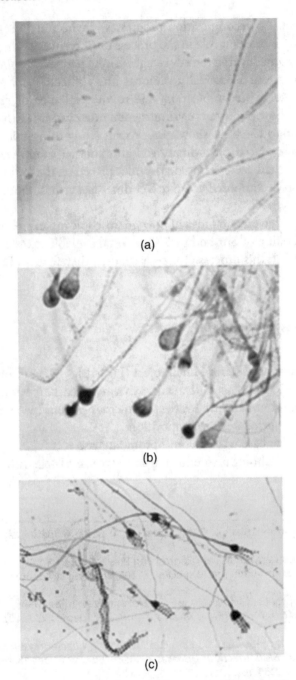

Figure 5.21 Microscopic appearance of some fungi that are frequently present in soil (a) *Geotrichum* showing long filaments as well as smaller, oblong spores (X300) Dr. Hilliard F. Hardin, Centers for Disease Control and Prevention, USA

An interdisciplinary review of the thanatomicrobiome in human decomposition Javan GT, Finley SJ, Tuomisto S, Hall A, Benbow ME, Mills D. *Forensic Science Medicine and Pathology* 2019;15(1):75-83. doi: https://doi.org/10.1007/s12024-018-0061-0.

Succession of oral microbiota community as a tool to estimate postmortem interval.
Dong K *et al*. *Scientific Reports* 2019;9(1):13063. doi: https://doi.org/10.1038/s41598-019-49338-z
https://pubmed.ncbi.nlm.nih.gov/31506511/

Potential use of bacterial community succession for estimating post-mortem interval as revealed by high-throughput sequencing. Guo J *et al*. *Scientific Reports* 2016; 6: 24197
https://www.ncbi.nlm.nih.gov/pmc/articles/PMC4823735/

The thanatomicrobiome: A missing piece of the microbial puzzle of death. Javan GT, Finley SJ, Abidin Z, Mulle JG. *Frontiers in Microbiology* 2016 Feb 24;7:225.
doi: https://doi.org/10.3389/fmicb.2016.00225.

Estimating the postmortem interval using microbes: Knowledge gaps and a path to technology adoption. Metcalf JL. *Forensic Science International: Genetics.* 2019 Jan;38:211-218. doi: https://doi.org/10.1016/j.fsigen.2018.11.004.
https://www.sciencedirect.com/science/article/pii/S1872497318304034

Thanatomicrobiome composition profiling as a tool for forensic investigation.
Zhou W, Bian Y. *Forensic Science Research* 2018;3(2):105-110. doi: https://doi.org/10.1080/20961790.2018.1466430. eCollection 2018. PMID: 30483658
https://www.ncbi.nlm.nih.gov/pmc/articles/PMC6197100/

Microbial forensics: new breakthroughs and future prospects. Oliveira M, Amorim A. *Applied Microbiology and Biotechnology.* 2018;102(24): 10377-10391. doi: https://doi.org/10.1007/s00253-018-9414-6. Epub 2018 Oct 9

(b). *Mucor* showing long filaments as well as large, onion-shaped, spore-containing structures (known as conidia) (x200)
Dr. Lucille K. Georg, Centers for Disease Control and Prevention, USA
(c). *Penicillium* showing long filaments as well as small, spherical spores (x475)
Dr. Lucille K. Georg, Centers for Disease Control and Prevention, USA

https://www.ncbi.nlm.nih.gov/pmc/articles/PMC7080133/

Human thanatomicrobiome succession and time since death. Javan GT, Finley SJ, Can I, Wilkinson JE, Hanson JD, Tarone AM. *Scientific Reports* 2016 Jul 14;6:29598. doi: https://doi.org/10.1038/srep29598
https://www.ncbi.nlm.nih.gov/pmc/articles/PMC4944132/

Cadaver thanatomicrobiome signatures: the ubiquitous nature of *Clostridium* species in human decomposition. Javan GT, Finley SJ, Smith T, Miller J, Wilkinson JE *Frontiers in Microbiology* 2017; 8:2096.
doi: https://doi.org/10.3389/fmicb.2017.02096
https://www.frontiersin.org/articles/10.3389/fmicb.2017.02096/full

Postmortem succession of gut microbial communities in deceased human subjects. DeBruyn JM, Hauther KA. *PeerJ.* 2017; 5: e3437.
https://www.ncbi.nlm.nih.gov/pmc/articles/PMC5470579/

Skin microbiome analysis for forensic human identification: what do we know so far? Tozzo P, D'Angiolella G, Brun P, Castagliuolo I, Gino S, Caenazzo L. *Microorganisms* 2020 Jun; 8(6): 873
https://www.ncbi.nlm.nih.gov/pmc/articles/PMC7356928/

Effects of extended postmortem interval on microbial communities in organs of the human cadaver. Lutz H *et al. Frontiers in Microbiology* 2020 Dec 8;11:569630. doi: https://doi.org/10.3389/fmicb.2020.569630. eCollection 2020. PMID: 33363519
https://www.ncbi.nlm.nih.gov/pmc/articles/PMC7752770/

The soil biology primer. US Department of Agriculture.
A useful introduction to the structure, biology and microbiology of soil
https://www.nrcs.usda.gov/wps/portal/nrcs/main/soils/health/biology/

Important microorganisms present in soil. Biology Discussion
https://www.biologydiscussion.com/soil-microbiology/8-important-microorganisms-present-in-soil-soil-microbiology/55510

Soil microbiology – a set of 36 teaching slides
https://www.slideshare.net/poojasabarinathan/soil-microbiology-53238336

Soil biology and microbiology. Andreas de Neergaard
In: land use, land cover and soil sciences – Vol. vi
http://www.eolss.net/Sample-Chapters/C12/E1-05-07-07.pdf

6

From the Micro to the Macro – Now the Big Guys Move In

Which insects are present on a human corpse has been of great interest to detectives and pathologists for many years because it helps them to establish the time of death of a deceased person. One of the earliest researchers in this area (a subject that is part of the wider field of "Forensic Entomology" – Box 6.1) was Jean Pierre Mégnin (1828–1905), a French veterinarian and entomologist. He published many articles and books on his research, especially on the succession of insects that occurs on corpses. One of the main conclusions from his research was that unburied corpses experienced eight waves of insect succession. This illustrates nicely what we'd expect following our studies of corpse microbiology – the environment of the corpse is largely responsible for dictating which insects can live there. In other words, the timing of the appearance of different insects on a corpse is another example of ecological succession. But how do insects know when a corpse is suitable for them? This is largely due to their ability to detect particular chemicals produced by the corpse at the different stages of decomposition. When the corpse is above ground, its appearance and colour are additional indications of its suitability for colonisation by a particular type of insect.

Interestingly, Megnin reported that when corpses were buried they experienced only four waves of succession. It's obvious how burial of a body would affect the insects living on a human corpse as this would prevent, or at least hinder, their access to it. Other factors that affect insect colonisation and succession include those listed in Table 4.4 in Chapter 4. This multiplicity of

M. Wilson, *Life After Death: What Happens to Your Body After You Die?*, Springer Praxis Books, https://doi.org/10.1007/978-3-030-83036-6_6

Box 6.1 Forensic entomology

Forensic entomology is a scientific field concerned with the study of insects, and other arthropods (i.e. invertebrates with a hard outer coating, segmented bodies and jointed appendages), with the purpose of providing evidence that will help legal investigations. Such information can be used to determine the time of death of a deceased individual, whether or not someone has been subjected to physical abuse, and is also of use in investigations of food contamination. In this book, we're concerned with the first of these applications and the insects that have proved to be the most useful for this purpose are blowflies, flesh flies, hide and skin beetles, rove beetles and clown beetles.

As we'll see later in this chapter, the various stages of corpse decomposition are associated with particular insects and other arthropods. Consequently, identification of which insects are present on a corpse, as well as what developmental stage they're in, enables an estimation of the PMI. However, it's important to remember that insect development is affected by a number of factors including climate and local weather conditions, so it's important that these are taken into account when interpreting findings. Temperature is of particular importance, so it's essential to determine what temperature the corpse has been kept at before any meaningful estimate of PMI can be made.

Among the insects of greatest value to determining the PMI are blowflies (family Calliphoridae) because they're usually the first insects to arrive on a corpse – often within hours of death. By determining the age of the oldest larvae it's possible to get an accurate estimate of the time of death.

factors does, of course, make it difficult to generalise about which insects appear at what PMI on a corpse. For reasons explained in Section 4.1, much of our understanding of the role of insects (as well as microbes) in corpse decomposition has come from studying corpses that have been left undisturbed out in the open air in research institutes.

An insect may be attracted to a corpse either because its flesh is an important food source or because it can feed on other insects it finds there. Broadly speaking it's possible to recognise four main types of insects associated with human corpses:

- those that feed directly on the corpse - known as "necrophagous species"
- predators of the insects that feed on the corpse
- those that feed on the corpse as well as on the insects found there - known as "omnivorous species"
- those that just use the corpse as a place to live – known as "adventive species"

An insect, unlike a human, undergoes a series of very dramatic changes during the course of its life. These changes (technically known as "metamorphoses") are so extensive that it's often very difficult to believe that the different creatures seen at each stage (larva, pupa) could possibly be the same organism. In order to appreciate the involvement of insects in corpse decomposition it's essential to know something about their life cycles. This is because, in many cases, it's not the fully-grown insect that's directly responsible for damage to the corpse, but its larvae.

6.1 What Is an Insect?

An insect is an animal with six legs and its body is divided into three parts – head, thorax and abdomen (Figure 6.1). Instead of having an internal skeleton, like us, it has an exoskeleton i.e. it's enclosed within a hard casing made of a polysaccharide known as chitin. On its head there are two jointed antennae (used as sense organs), a pair of eyes and a set of mouthparts. Its thorax has three sections and attached to each is a pair of jointed legs. The thorax also has one or two pairs of wings, depending on the particular insect.

Insects feed in a variety of ways and have specialised mouthparts designed to accomplish this. Some (e.g. beetles, bees, wasps, ants) have powerful mandibles that enable them to chew food and then digest it internally (Figure 6.2a). Others (e.g. flies) secrete enzyme-containing saliva onto the food (Figure 6.2b) and then suck up the resulting digested food through a sponge-like structure.

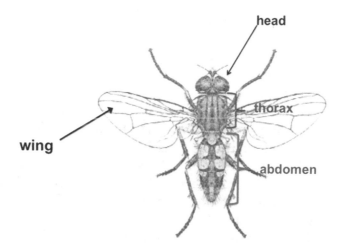

Figure 6.1 Diagram showing the main distinguishing features of an insect

(a)

(b)   (c)

Figure 6.2   The three main types of mouthparts found in insects
(a) The large and powerful mandibles on the head of a ground beetle
Whitney Cranshaw, Colorado State University, Bugwood.org
Creative Commons Attribution 3.0 License.
(b) The sponge-like mouthparts of a fly belonging to the genus *Sarcophaga*.
Author: Siga. Permission is granted to copy, distribute and/or modify this document under the terms of the GNU Free Documentation License, Version 1.2 or any later version published by the Free Software Foundation; with no Invariant Sections, no Front-Cover Texts, and no Back-Cover Texts.
(c) The Convolvulus hawk moth with its long proboscis
Сделал сам, Public domain, via Wikimedia Commons

Moths and butterflies have a proboscis (Figure 6.2c) which enables them to suck up liquid food such as nectar.

As mentioned previously, an insect has a complex life cycle during which it undergoes a series of dramatic changes (Figure 6.3). In most insects, reproduction involves mating between a male and female after which the female lays eggs. An egg hatches into a larva which usually moves around actively as it grows and undergoes a series of moulting stages until it eventually becomes a pupa. A pupa shows little movement and is often enclosed within a cocoon or puparium. Within this protective structure the pupa undergoes a process of metamorphosis and turns into the adult insect which then emerges.

If we apply scientific terminology, an insect is described as being an animal that belongs to the Insecta class within the Arthropoda phylum. Approximately three quarters of all the known animals on Earth are insects and it's been

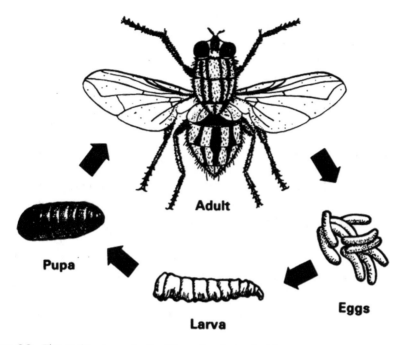

Figure 6.3   The major stages in the life cycle of a typical insect
Pratt HD, Littig KS, Scott HG. Center for Disease Control, Public domain, via Wikimedia Commons

estimated that there may be as many as 10 million different insect species. Within the Insecta class there are 29 different Orders. The Orders with the greatest number of species are Coleoptera (beetles), Lepidoptera (butterflies and moths), Hymenoptera (ants, bees and wasps) and Diptera (true flies).

6.2   What Goes on Above Ground – Who's First and Who's Last?

Just as with microbes, much of what we know about insect succession on a human corpse comes from studying bodies that have been left unburied. These may have been left intentionally in the open air by scientists researching this phenomenon, by criminals disposing of their victims or by people dying in sparsely-populated regions. The succession of insects found on a corpse is affected by a large number of factors including:

- the nature of the climatic region (tropical, temperate etc.)
- the season

- the weather at the time
- the type of environment i.e. whether it is rural or urban
- the presence of clothing
- whether or not the corpse is wrapped in some material

Although we're going to try to get some idea of the succession of insects that colonise a human corpse above ground, it's important to realise that this isn't straightforward. It's not as if insect A starts off on a new corpse and then insect B comes along and A simply disappears. It's not at all like that. There's tremendous overlap of species at the various stages of decomposition. Each stage usually attracts many species and some of these disappear by the next stage but most persist to the next and, often, to subsequent stages. Table 6.1 will give you some idea of just how complicated it is. This shows the number of species that have been found on human corpses at each stage of decomposition and also the percentage of these species that are also found at other stages.

So, what does this table tell us? First of all you can see that the number of insect species increases as decomposition proceeds until it reaches a maximum (426 species) during the advanced decay stage – it then starts to decrease. Another obvious thing is that most (94%) of the species attracted at the fresh stage persist on the corpse through to the active stage. Few of the insects that are active in the early stages (fresh, bloat and active decay) remain on the corpse until the skeletonization stage. Also, it shows that most of the species active at the skeletonization stage arrive on the corpse late in the decomposition process, mainly (76%) in the advanced decay stage.

So, let's now take a look at which insects are found on the corpse at various stages in its decomposition.

Table 6.1 Insect species at each of the decomposition stages of a human corpse

Decomposition stage	Number of species at that decomposition stage	Percentage of species attracted to another decomposition stage				
		Fresh	Bloat	Active decay	Advanced decay	Skeletonisation
Fresh	17	100	94	94	76	0
Bloat	48	33	100	100	90	2
Active decay	255	6	19	100	98	13
Advanced decay	426	3	10	59	100	38
Skeletonisation	211	0	1	16	76	100

6.2.1   The Fresh Stage

Most studies agree that the first insect to arrive at a corpse that's been left above ground is the blow fly (Figure 6.4) - also known as the bluebottle, greenbottle and carrion fly. It's doubtful, however, that blow flies can find corpses buried more than 20 cm below the surface.

Blow fly is the common name for the insect family Calliphoridae which consists of 1,100 species in 62 genera. The most frequently-encountered species include *Calliphora vicina* (bluebottle), *Calliphora vomitoria, Phormia regina* (black blow fly) and *Lucilia sericata* (greenbottle) – the particular

(a)   (b)

(c)   (d)

Figure 6.4   Blow flies
(a) Blow fly
gailhampshire from Cradley, Malvern, U.K, CC BY 2.0 <https://creativecommons.org/licenses/by/2.0>, via Wikimedia Commons
(b) Drawing of a greenbottle (*Lucilia sericata*), a type of blow fly.
Image courtesy of the Centres for Disease Control and Prevention, USA
(c) Drawing of a black blow fly (*Phormia regina*)
Image courtesy of the Centres for Disease Control and Prevention, USA
(d) Drawing of a bluebottle (*Calliphora vicina*)
Image courtesy of the Centres for Disease Control and Prevention, USA

species vary according to the climatic region. Adult flies are 8-10 mm long and generally feed on nectar and are important pollinators of flowers.

The blow fly is attracted to a corpse within minutes of death because the corpse produces volatile compounds such as dimethyl sulphide and dimethyl trisulfide. Amazingly, a blow fly can, under ideal conditions, detect such compounds when it's several miles away. However, they're not attracted to corpses that are in the later stages of decomposition – especially the dry and skeletonisation stages. They are active during the day and so a corpse that's been deposited at night isn't likely to be colonised by them until the following day. Blow flies don't actually feed on the corpse but they lay eggs in the orifices (mouth, nose, ears, anus, eyes) or wounds of the corpse because they can't break through its skin. A single fly can lay as many as 250 eggs each of which is about 1.2 mm long and 0.3 mm wide and is white and shiny. These then hatch into larvae usually within about 24 hours (Figure 6.5). The larvae of all fly species are also known as maggots.

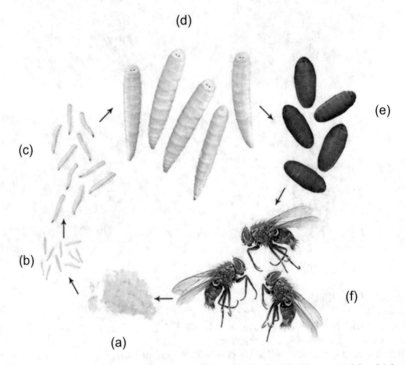

Figure 6.5 Life cycle of *Calliphora vicina*. The adult fly (f) lays eggs (a) which then hatch to form maggots (b). These grow and moult (c, d) to form a puparium (e) in which metamorphosis occurs to produce the adult fly (f)
Image courtesy of the Natural History Museum, London © The Trustees of the Natural History Museum, London. Licensed under the Open Government Licence

The maggot is white and consists of 12 segments with a black, pointed front end that has jaws (mandibles) and hooks. Fluids containing digestive enzymes are secreted both from the front and back ends of the maggot and these can break down the macromolecules present in the tissue of the corpse – the nutrients are then sucked up by the maggot through its mouth. The maggot then grows and passes through two moulting stages and by then it's often about 2 cm long. It then moves away from the corpse and seeks out a dark and cool place where it becomes a pupa covered in a hard skin (cuticle) known as a puparium. This migration of larvae away from the corpse is a common phenomenon among many species that feed on corpses. Often the larva buries itself in neighbouring soil (this can be up to 30 m away) to hide away as protection during this static, pupation stage during which it's obviously very vulnerable to predators. The puparium changes colour and eventually becomes dark red or brown. Inside the puparium a remarkable transformation takes place resulting in an adult fly. The timing of these various stages depends very much on temperature and when this is about 20°C the overall process from egg laying to the emergence of an adult fly takes about 2 weeks (Table 6.2).

As well as consuming readily-available nutrients such as sugars and amino acids present in the corpse, maggots secrete a range of macromolecule-degrading enzymes that supply them with even more nutrients. These enzymes include lipases, nucleases, glycosidases and a wide range of proteases. Some of the proteases secreted are specifically aimed at breaking down important extracellular tissue proteins such as collagen, fibronectin, elastin and laminin which are responsible for maintaining the structure of human tissues. So, it's the maggots that feed on the corpse that help to bring about its decomposition rather than the adult flies. This has led to the frequently-quoted observation made by Carl Linnaeus (the founder of the two-part naming system used for all forms of life) in 1767 that "three flies could consume a horse cadaver as rapidly as a lion".

Often the maggots are present in huge numbers (hundreds or thousands) that congregate together in a "maggot mass" (Figure 6.6). A maggot mass may

Table 6.2 The timings of the various stages in the life cycle of a blow fly at 20°C

Time (hours)	Event
0	Adult fly lays eggs
23	Eggs hatch
50	Maggots moult
72	2nd moulting
202	Formation of a puparium
345	Adult fly emerges

Figure 6.6   A maggot mass formed by blow fly larvae
Susan Ellis, Bugwood.org. Licensed under a Creative Commons Attribution 3.0 License.

consist of the larvae of one species but, more usually, the larvae of several species are present. The movement and metabolism of such a huge number of living creatures produces a lot of heat and can result in temperatures 30°C higher than that of the environment. Why exactly maggots form such large aggregates isn't really understood. It could be the result of eggs being deposited in large groups and/or mutual attraction resulting from chemical signals (pheromones) produced by the larvae. It's important to bear in mind that these maggot masses may not appear until the bloat or active decay stages.

The ability of maggots to break down human tissues has been used in an interesting approach to the treatment of wounds (Box 6.2).

Box 6.2 Maggot therapy for wounds

Maggots have been used to treat wounds for many centuries – both the Mayans and native Australians successfully practiced this approach. Then, during the American Civil War, it was noticed that maggots in wounds not only removed dead tissue but also promoted wound healing. However, interest in this approach decreased with the discovery, and wide-scale use, of antibiotics in the middle of the 20th century. Nevertheless, the increased resistance of many bacteria to antibiotics has stimulated considerable interest in using maggots for treating wounds and several clinical trials have shown them to be highly effective.

Box 6.2 (continued)

So, what do maggots actually do in a wound? An important aspect of wound treatment is the removal of any dead tissue that may be present – this is known as debridement. Traditionally, debridement is carried out using sharp instruments and this is known as "surgical debridement". Maggots, as we know, feed on dead tissue and so can be used to do this – they eat the dead tissues present in a wound without damaging the live tissue. Also, they produce ammonia (which kills microbes) and their secretions contain antimicrobial compounds that kill any microbes that may be infecting the wound. Furthermore, they eat microbes. The presence of the maggots also stimulates the immune response of the patient. They therefore have a triple action – removal of dead tissue, disposal of any microbes that are present and stimulation of the immune defence system.

In practice, sterilised maggots of the green bottle fly, *Lucilia sericata* (Figure a) are added to the wound which is then covered with a dressing to keep them in place. They're left there for about 2 days, and during this time they grow in size from about 7 mm to 13 mm. The dressing is then removed, and the maggots wiped away with a wet gauze and sterile saline. It's important that only sterilised maggots are used, otherwise they would be introducing microbes into the wound which could cause an infection. This is achieved by treating the eggs with sodium hypochlorite, the maggots are then hatched under sterile conditions.

(a)

Figure (a) *Lucilia sericata*
Image courtesy of wild_wind and the Global Biodiversity Information Facility. CC BY 4.0.

(b)

Figure (b) Drawing of the larva of *Lucilia sericata*
Coloured drawing by A.J.E. Terzi. Credit: Wellcome Collection. Attribution 4.0 International (CC BY 4.0)

Although the adult flies can arrive on a corpse during the fresh stage, it may be several days later before the maggots they produce actually become involved in its decomposition. Their effects, therefore, may not become apparent until later stages such as during bloat and active decay.

Sarcophagidae (also known as flesh flies) may also start to arrive during the fresh stage although their presence is generally more obvious during the bloat stage. Their name comes from the Greek "sarx" and "phagein" meaning "flesh" and "to eat" respectively. They tend to arrive concurrently with, or slightly after, blow flies. They are between 4 and 23 mm long and are similar to blow flies in appearance, but they are blackish and have grey longitudinal stripes on the thorax (Figure 6.7). Worldwide, there are about 3,000 species in more than 170 genera. The adult flies feed on nectar, excrement as well as the flesh of, and fluids from, the bodies of animals. They differ from most other types of fly in that, following mating, the eggs hatch within the female's body. On arriving at a corpse, the female deposits her larvae directly onto it. The larvae feed on the corpse for 4 - 7 days and, after two moulting stages, reach a size of about 9 – 13 mm and then form pupae. The pupa is 5-10 mm long and an adult emerges from it after 10 - 14 days.

As well as consuming the tissues of the corpse, flesh flies also contribute to its decomposition in another way – they deposit microbes onto it. All of the other insects that land on a corpse will, of course, also do this. However, because most of the microbes on insects are also found in the environment or on humans, we can't be certain which microbes present on a corpse have come specifically from insects. The exceptions to this are species belonging to the genera *Ignatzschineria* and *Wohlfahrtiimonas* (Box 6.3) which are found mainly in the larvae of various flesh flies. In chapter 5 we saw that these bacteria can comprise a significant proportion of the microbiota of various body sites during corpse decomposition.

6.2.2   The Bloat Stage

During this stage a variety of beetles are found on the corpse, including members of the following families: Staphylinidae (rove beetles), Silphidae (carrion beetles) and Histeridae (clown beetles). Blow flies are still present and are usually joined by the muscid fly *Hydrotaea capensis* (also known as *Ophyra capensis*). Flesh flies are also very active during the bloat stage, although they may have arrived during the fresh stage.

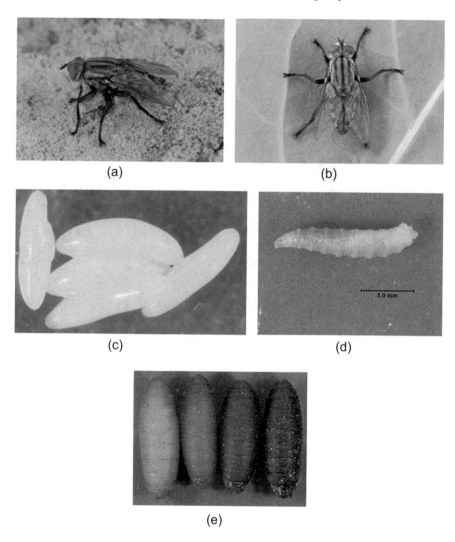

(a)

(b)

(c)

(d)

5.0 mm

(e)

Figure 6.7   (a) and (b) Adult *Sarcophaga crassipalpis* (flesh fly). (c) eggs dissected from a female before they had hatched (d) larva (e) pupae showing the colour changes that occur as they age
(a) and (b) Images courtesy of Lyle Buss, University of Florida
(c), (d) and (e). Images courtesy of Lazaro A. Diaz, University of Florida

Box 6.3 Bacteria originating from insects

Ignatzschineria are Gram-negative, non-motile aerobic bacilli that live in the gut of the larvae of the parasitic wolf fly (*Wohlfahrtia magnifica*) as well as in the gut of adult flesh flies (*Sarcophaga* species). They can grow at temperatures as

Box 6.3 (continued)

low as 4°C and so can grow on the corpse once it reaches the low temperatures found in the environment. They can break down lipids and peptides.

Wohlfahrtiimonas are Gram-negative, aerobic bacilli (Figure) that are non-motile. They were first isolated from the larvae of the fly *Wohlfahrtia magnifica*. They can hydrolyse polysaccharides including glycogen which is the main energy storage polysaccharide in humans and other animals.

Figure. Gram stain of *Wohlfahrtiimonas chitiniclastica* **showing Gram-negative bacilli**
Image courtesy of Belinda Lin and Owen Harris (2016), St John of God Pathology, Geelong, Victoria, Australia

A number of other flies visit a corpse during the early stages of its decomposition without laying their eggs there and these include species belonging to the genera *Pollenia*, *Ravinia* and *Oxysarcodexia*. They feed off the corpse itself or fluids from it.

Rove beetles (Staphylinidae – Figure 6.8), which are predators that feed on the eggs and maggots of flies, are usually present. Although they don't actually feed on the corpse itself, they can affect its rate of decomposition by reducing the number of maggots. Staphylinidae are the largest family of organisms on Earth – there are more than 63,000 species in 3,200 genera. The adults are generally long and slender and vary in size from 1 – 25 mm. The larvae are pale-coloured, long and slender and, like the adults, have mandibles for chewing and they feed on the larvae and eggs of other insects.

The Silphidae family of beetles are also known as carrion or burying beetles and consist of 1500 species in 15 genera. They are between 7 and 45 mm long, generally dark in colour and have flattish bodies. Their larvae also tend

Figure 6.8   Example of a rove beetle - *Platydracus maculosus*
Susan Ellis, USDA APHIS PPQ, Bugwood.org. Licensed under a Creative Commons
Attribution 3.0 License

to be flattened and are 15 – 30 mm long. Frequently-encountered genera
include *Necrophila* (Figure 6.9a), *Nicrophorus* (Figure 6.9b) and *Thanatophilus*
(Figure 6.9c). However, *Nicrophorus* species (also known as sexton or burying
beetles) generally prefer the corpses of small animals such as mice which they
bury (Box 6.3).

The life cycle of these beetles involves the sequence of events illustrated in
Figure 6.10.

Both the adult beetles and their larvae (Figure 6.11) feed on the tissues of
the corpse using their strong mandibles to chew the flesh which is then
digested internally. They are most frequently present during the bloat and
active decay stages.

The Histeridae family, also known as clown beetles, are small (1-10 mm
long), oval-shaped, shiny, beetles with a leathery texture. There are 4,800 species in 410 genera. Species belonging to the genera *Hister* (Figure 6.12a) and
Saprinus (Figure 6.12b) are frequently found on human corpses. They tend to
stay underneath the corpse during the day and become active at night.

They take between 1 and 2 months to develop from the egg to the adult
clown beetle, passing through two larval stages and pupation. Both the adults
and larvae feed mainly on the larvae and eggs of flies that have colonised the
corpse. When touched, they pretend to be dead - this phenomenon is known
as thanatosis and is probably a defence mechanism. They are generally present
mainly during the bloat and active decay stages.

(a)

(b)   (c)

Figure 6.9 Examples of beetles belonging to the Silphidae family (a) *Necrophila americana,* (b) *Nicrophorus vestigator* (c). *Thanatophilus trituberculatus*
(a) Smithsonian Environmental Research Center, CC BY 2.0 <https://creativecommons.org/licenses/by/2.0>, via Wikimedia Commons
(b) Stanislav Snäll, CC BY 3.0 <https://creativecommons.org/licenses/by/3.0>, via Wikimedia Commons
(c) Stanislav Snäll, CC BY 3.0 <https://creativecommons.org/licenses/by/3.0>, via Wikimedia Commons

Box 6.3 What wonderful parents

Nicrophorus species (commonly known as "burying beetles") are unusual among beetles because they actually take care of their offspring. Most species breed on the corpses of small animals such as mice and birds which they bury in holes that they dig specifically for this purpose – they do this by pushing soil around with their heads. Before burying a corpse, they remove its fur or feathers (and use these to line the burial chamber) and smear it with antimicrobial chemicals that slow its decay and make it less attractive to other insects. The eggs are laid in nearby soil and the larvae then move into the chamber. The larvae beg for food from their parents who respond by regurgitating liquid food that they've obtained from the corpse. They do this for about 7 days before flying off, leaving their well-fed offspring behind.

Figure. *Nicrophorus vespilloides* **on the corpse of a rat**
Vexillum, CC BY 3.0 <https://creativecommons.org/licenses/by/3.0>, via Wikimedia Commons

Muscid flies belong to the family Muscidae which consists of more than 9000 species in over 190 genera. They are 3 – 10 mm long, generally grey in colour and have lines along their thorax. The family includes the common house fly, *Musca domestica*, (Figure 6.13a) and the face fly, *Musca autumnalis* (Figure 6.13b), both of which are found on human corpses. The life cycle of the domestic fly is shown in Figure 6.14. Rather than feeding on the corpse itself, the adults flies feed on faeces and any fluids that have exuded from the corpse such as urine, saliva, blood etc. Feeding in adults involves secretion of enzyme-containing saliva which breaks down macromolecules, the resulting nutrients are then sucked up.

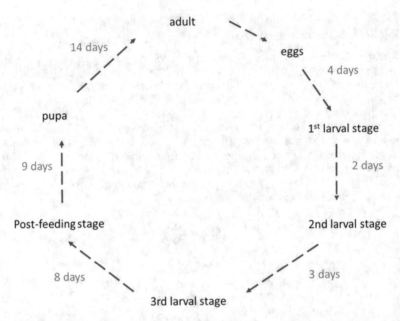

Figure 6.10   Life cycle of *Necrophila* species – this is typical of other members of the Silphidae family of beetles. The timings shown are for a temperature of 20°C

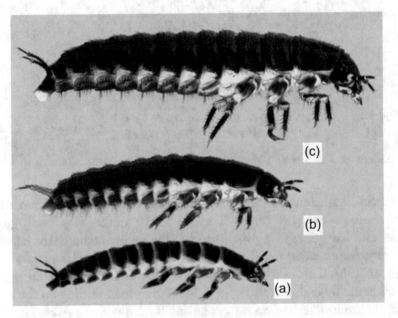

Figure 6.11   First (a), second (b) and third (c). stage larvae of *Thanatophilus rugosus*. The first stage larva is approximately 0.7 cm long

Image from: Revisited larval morphology of *Thanatophilus rugosus* (Coleoptera: Silphidae). Novák M *et al. International Journal of Legal Medicine* 132, 939–954 (2018). https://doi.org/10.1007/s00414-017-1764-6

(a)   (b)

Figure 6.12   Examples of clown beetles. (a) *Hister unicolor*, (b) *Saprinus maculatus*
(a) Image courtesy of UK Beetles, Creative Commons Attribution 4.0 International License
(b) Shayya S, Dégallier N, Nel A, Azar D, Lackner T (2018) Contribution to the knowledge of Saprinus Erichson, 1834 of forensic relevance from Lebanon (Coleoptera,
Histeridae). *ZooKeys* 738: 117-152. https://doi.org/10.3897/zookeys.738.21382, CC BY
4.0 <https://creativecommons.org/licenses/by/4.0>, via Wikimedia Commons

Eggs are laid on the corpse and these hatch into larvae which feed on the
corpse. The larvae are 5 – 12 mm long, slender, white or cream, headless and
taper towards the front end (Figure 6.13a). Instead of a head, a larva has a pair
of mouth-hooks which it uses to scrape off food into its mouth.

Hydrotaea (or *Ophyra*) species (Figure 6.15) are another type of muscid fly
attracted to corpses at this stage. They are light brown to bluish black in
colour with large, red eyes and are between 6 and 9 mm long. The adults often
feed on mammalian blood while the larvae feed on tissues and the larvae of
other, or their own, species.

During this stage maggot masses are often visible on the corpse. These
could consist of the larvae of species that laid their eggs during the fresh stage
of decay and/or those that hatched during this bloat stage.

6.2.3   The Active Decay Stage

Many of the insects mentioned as appearing on a corpse during the bloat stage
also persist into the active decay stage. Larvae from these earlier visitors may
form maggot masses during this stage. New visitors, particularly towards the
end of this stage, include members of the Nitidulidae family (sap beetles).

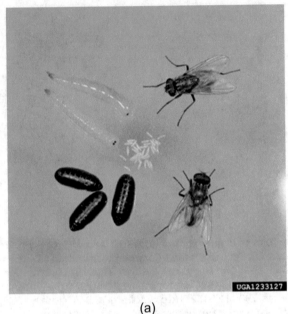

(a)

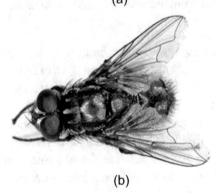

(b)

Figure 6.13 (a). Adults, eggs, larvae and pupae of the common house fly, *Musca domestica*. (b) The face fly, *Musca autumnalis*
(a) Clemson University - USDA Cooperative Extension Slide Series , Bugwood.org. Creative Commons Attribution 3.0 License
(b) This image is created by user B. Schoenmakers at waarneming.nl, a source of nature observations in the Netherlands., CC BY 3.0 <https://creativecommons.org/licenses/by/3.0>, via Wikimedia Commons

Sap beetles comprise more than 2500 species. They are 4 – 12 mm long with clubbed antennae and are generally dark in colour (Figure 6.16). Although many species feed on sap and over-ripe fruit, species of the genera *Omosita* and *Nitidula* are found on corpses during the later stages of decomposition. Their larvae are white or pale yellow with a light brown head and three pairs of legs.

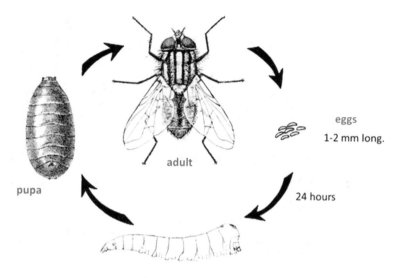

eggs
1-2 mm long.

adult

pupa

24 hours

larva   1-2 mm long.

Figure 6.14 The life cycle of the house fly
Unknown author, Public domain, via Wikimedia Commons

Figure 6.15 *Hydrotaea* (or *Ophyra*) species
S. Rae from Scotland, UK, CC BY 2.0 <https://creativecommons.org/licenses/by/2.0>, via
Wikimedia Commons

Figure 6.16 Images of sap beetles. (a) *Omosita colon* (b) *Nitidula* species (c) larva of a sap beetle
(a) Susan Ellis, Bugwood.org. licensed under a Creative Commons Attribution 3.0 License.
(b) Georgiy Jacobson, Public domain, via Wikimedia Commons
(c) Jeffrey W. Lotz, Florida Department of Agriculture and Consumer Services, Bugwood. org. Licensed under a Creative Commons Attribution 3.0 License.

6.2.4   The Advanced Decay Stage

As the corpse dries out, it becomes less suitable for blowflies, flesh flies and house flies which all prefer a semi-liquid environment. This results in a decrease in the fly population during this stage while beetles often remain numerous. In addition to the beetles mentioned as being present during active decay, members of the families Dermestidae, Cleridae (checkered beetles) and Scarabaeidae (lamellicorn beetles) are also frequently present.

Members of the Dermestidae family are also known as skin, larder, hide, leather and carpet beetles – the names giving an idea of their preferred food. Nearly all of these are able to feed on dried animal tissues. Species belonging to the genus *Dermestes* (skin beetles) are frequently found during the advanced decay and skeletonisation stages of corpse decomposition. They've been reported to be able to reduce a human corpse to a skeleton in only 24 days. This flesh-removing ability has been taken advantage of for many years to remove flesh from museum specimens (Box 6.4).

Box 6.4 Cleaning beetles

Many museums use flesh-eating beetles to help remove any flesh present on their skeletons before they're put on display. *Dermestes* species are the most widely used for this purpose and can remove any flesh relatively quickly. *Dermestes maculatus* is the species most widely used. A great advantage of using beetles is that, unlike mechanical cleaning methods, they don't do any damage to delicate parts of the skeleton. Chemicals used to clean skeletons, such as hydrogen peroxide and carbon tetrachloride, can also do a lot of damage because they tend to make bones more fragile. As well as being used to clean skeletons of any residual flesh, museums can also use *Dermestes* beetles to produce skeletons from entire dead animals (an example of this can be seen in Appendix III). The beetles can convert small animals such as mice to a skeleton overnight.

Dermestes beetles are 3-12 mm long with an oval shape and dark colour (black or dark brown) and have light-coloured spots (Figure 6.17a). Mature larvae are generally brownish in colour, 11–13 mm in length, have chewing mouthparts and are covered with strong bristle-like hairs of different sizes (Figure 6.17b). Importantly, they can digest keratin, the main protein of skin and hair. The most commonly reported, and widely distributed, species is *D. maculatus* and its life cycle is summarised in Figure 6.18. There's considerable interest in using this species to help determine the PMI. Generally the beetle arrives during the later stages of decomposition when only skin and bone remain, about 3-6 months after death. But it's been found as early as 3-11 days after death, although larvae weren't present until the dry decay or skeletonisation stages. Environmental, particularly climatic, factors strongly influence the arrival time and activity of the beetle on a corpse.

Checkered beetles (Cleridae) are usually brightly-coloured with elongated bodies covered in bristly hairs and are 3 – 24 mm in length (Figure 6.19). There are more than 3500 species and the adults feed on the skin of the corpse

Figure 6.17 Images of *Dermestes* beetles. (a) adult, (b) larvae, (c) pupa
(a) UK Beetles, Attribution 4.0 International (CC BY 4.0)
(b) Whitney Cranshaw, Colorado State University, Bugwood.org. Creative Commons License licensed under a Creative Commons Attribution 3.0 License.
(c). JaigoL, CC BY 4.0 <https://creativecommons.org/licenses/by/4.0>, via Wikimedia Commons

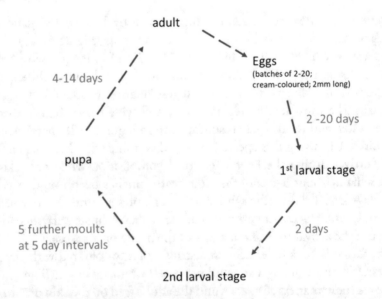

Figure 6.18 Life cycle of *Dermestes maculatus*

(a)

(b)

Figure 6.19   An adult checkered beetle (*Thanasimus dubius*) (a) and its larva (b)
Gerald J. Lenhard, Louisiana State University, Bugwood.org. licensed under a Creative
Commons Attribution 3.0 License.

as well as on other insects. The larvae generally feed on the larvae of other
insects but some species can devour flesh. They prefer the drier conditions
found during the later stages of decomposition

The Scarabaeidae (lamellicorn beetles) are a huge family with over 19,000
species which display a variety of shapes, sizes and colours. A sub-family,
known as dung beetles (Figure 6.20a) or tumble bugs, live on dung (Box 6.5)
and corpses. These are dull-coloured and rounded and generally less than
2.5 cm long. The larvae (Figure 6.20b) are C-shaped and white with a
brown head.

6.2.5   The Skeletonisation Stage

At this stage, all that remains of the corpse are bones and hair. Various beetles
are often found, including the skin, sap, checkered and lamellicorn beetles
described previously. In addition, Trogidae (trogid or hide beetles) are often

(a)

(b)

Figure 6.20 Example of a dung beetle (a) and its larva (b)
(a) Moretto Philippe, Perissinotto Renzo, CC BY 3.0 <https://creativecommons.org/licenses/by/3.0>, via Wikimedia Commons
(b). Clemson University - USDA Cooperative Extension Slide Series , Bugwood.org licensed under a Creative Commons Attribution 3.0 License.

Box 6.5 Using stars as a navigation system

The nocturnal African dung beetle *Scarabaeus satyrus* is one of the few non-vertebrate animals that can navigate and orient themselves using the Milky Way. Once a dung beetle has found a pile of dung it takes a small portion and makes a small ball of it. It then needs to quickly move the dung ball away to stop it being stolen by the other beetles that will have congregated around the pile. If it just randomly pushed its dung ball away, then it might just end up back near the pile and have it stolen. It's important, therefore, that it knows how to move away in a straight line, and so needs to be able to orientate itself. A series of experiments carried out in the Johannesburg Planetarium showed that the beetles use light-receptors in their heads to detect light from the Milky Way and found that this could be used for orientation purposes. As long as they could detect the Milky Way, the beetles could move in a straight line. If their heads were covered by cardboard shields, they moved around randomly. Other species,

Box 6.5 (continued)

such as *Scarabaeus zambesianus,* use the moon to navigate in straight lines, but just wander around randomly on moonless nights.

(a)

Figure (a) Dung beetles on a pile of horse dung
Duwwel, Public domain, via Wikimedia Commons

As well as being great navigators, dung beetles are also very strong. One species, *Onthophagus taurus,* is thought to be the strongest animal on earth because it can move up to 1141 times its body weight.

(b)

Figure (b) *Onthophagus taurus*
Image courtesy of Gabriele Vaudano and the Global Biodiversity Information Facility. CC BY 4.0.

Figure 6.21 Trogid beetle
LiCheng Shih, CC BY 2.0 <https://creativecommons.org/licenses/by/2.0>, via Wikimedia Commons

present – this is a small family consisting of about 300 species. Trogids (Figure 6.21) are oblong or oval in shape and are 5-20 mm long. They are usually black, brown or grey in colour and have a rough, bumpy appearance. They're often covered in soil or other debris. Their larvae are C-shaped, white or cream-coloured, and feed on skin, hair and any dried tissue that has stuck to bones.

Moths that usually feed on our clothes sometimes lay eggs on the corpse after the flies have left and these hatch out and the larvae then feed on any remaining hair. These moths belong to the Family Tineidae and include *Tineola bisselliella*, also known as the "common clothes moth" (Figure 6.22).

6.2.6 Wave Upon Wave of Insects – "Help, I'm Not Waving But Drowning"

Feeling overwhelmed? Let's summarise all the above information so you can have a "take home message". The above description shows that a human corpse is exposed to a series of invasions by different insects. This will remind you of what happens to a corpse when studied from the point of view of microbes. In each case, a succession of different species occurs and this can be explained in ecological terms – basically the original species (whether on the

Figure 6.22   The common clothes moth, *Tineola bisselliella*. A larva (bottom left), pupa (centre left) and an adult (centre right) are shown
Clemson University - USDA Cooperative Extension Slide Series , Bugwood.org
licensed under a Creative Commons Attribution 3.0 License

micro- or macro- scale) alters the environment of the corpse and paves the way for other species which, in turn, alter the environment again. A summary of the insect succession that takes place in an unburied human corpse is given in Figure 6.23. This gives an idea of when particular adult insects are likely to be found on a corpse.

Figure 6.24 shows the sequence of appearance of the larval form of the insects on a corpse – these are often responsible for most of the decomposition that occurs.

Table 6.3 summarises which insects are active during the various stages of corpse decomposition as well as what they do to the corpse.

6.3   And Now for What's Going on Down Below

As explained in Chapter 4, it's far more difficult to study buried corpses than those that are left above ground. In general, fewer species of insects are associated with corpses that are buried compared to those that are left above ground and even a thin layer of soil has a significant effect on which insects are present. Also, the sequence of colonisation is different. These factors (together

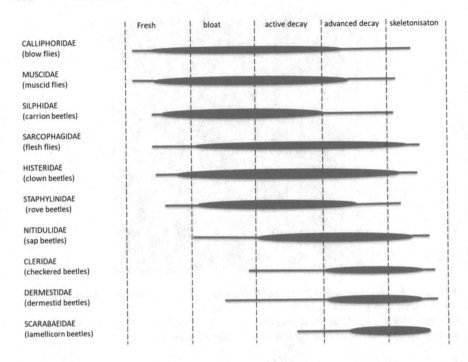

Figure 6.23 The succession of adult insects on an unburied human corpse. This is a general summary and the exact sequence is influenced by a number of factors including the nature of the corpse, the climate and the weather. The width of the oval shapes gives an indication of the number of insects present at a particular decomposition stage. The horizontal lines indicate the range of decomposition stages over which the insects are likely to be found

with the effect that burial has on the types of microbes present) combine to make decomposition a much slower process.

6.3.1 Early Arrivals

Blowflies, which play a major role in decomposing corpses above ground are generally excluded from buried corpses – even by a layer of soil only 2.5 cm thick. Some flies lay their eggs on the surface and their larvae then burrow down to feed on the corpse. Examples include flies belonging to the families Muscidae and Heleomyzidae. Other insects burrow down to the corpse and lay their eggs on it – these include members of the families Rhizophagidae, Staphylinidae, Phoridae, Braconidae and Proctotrupidae.

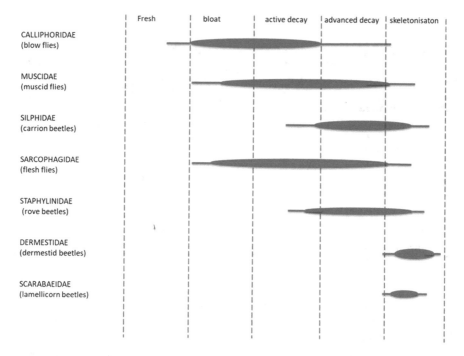

Figure 6.24 The succession of insect larvae on an unburied human corpse. This is a general summary and the exact sequence is influenced by a number of factors including the nature of the corpse, the climate and the weather. The width of the oval shapes gives an indication of the number of insects present at a particular decomposition stage. The horizontal lines indicate the range of decomposition stages over which the insects are likely to be found

Table 6.3 A summary of which insects appear during the various stages of corpse decomposition

Insect	Main decomposition stage(s) of activity	Main role in decomposition
blow flies	Fresh to active decay	Larvae feed on tissues
flesh flies	Bloat to advanced decay	Larvae feed on tissues
rove beetles	Bloat to active decay	Feed on larvae and eggs of other insects
carrion beetles	Fresh to active decay	Feed on tissue
clown beetles	Bloat to advanced decay	Feed on larvae and eggs of other insects
muscid flies	Fresh to active decay	Feed on tissue
sap beetles	Active decay to advanced decay	Feed on tissue
skin beetles	Advanced decay to skeletonisation	Feed on tissue
checkered beetles	advanced decay to skeletonisation	feed on larvae and eggs of other insects
dung beetles	Advanced decay to skeletonisation	Feed on tissue
hide beetles	Skeletonization	Feed on tissue

Table 6.4 Insect species found on human corpses at different burial times

Insect	Time since burial (days)			
	0-100	100-200	200-300	>300
Ophyra (Hydrotaea) capensis	+	+	+	+
Conicera tibialis	+	+	+	-
Leptocera caenosa	+	+	+	-
Triphleba hyalinata	+	+	-	-
Calliphora vicina	-	+	-	-
Megaselia rufipes	-	+	+	-
Omalium rivulare	-	+	-	-
Mites	-	-	+	-
Fannia scalaris	-	-	+	+
Fannia manicata	-	-	+	-
Philonthus species	-	-	-	+

In general, species belonging to the families Phoridae, Muscidae and Sarcophagidae are the dominant ones present on buried corpses. Unfortunately, because of the practical and ethical difficulties outlined previously (Section 4.1), we know little about insect succession on buried corpses. However, as mentioned at the start of this chapter, the famous entomologist, Pierre Mégnin, concluded that there were four waves of insect succession on buried corpses. Table 6.4 shows the results of one of the most extensive studies that has been undertaken so far.

Early colonisers of buried corpses include *Ophyra (Hydrotaea) capensis, Conicera tibialis, Leptocera caenosa,* and *Triphleba hyalinata.*

Ophyra species have been described previously in this chapter (Section 6.2.2). As well as being early arrivals, they appear to persist throughout the whole process of corpse decomposition.

Conicera tibialis is also known as the "coffin fly" (and, confusingly, so are several other fly species) because of its frequent presence in coffins. It's a member of the Phoridae family (also known as "scuttle flies") which consists of about 4,000 species in 230 genera. All of these are small (0.5-6 mm long), hump-backed flies which are usually black or brown (Figure 6.25). *Conicera tibialis* is 1.5 – 2.5 mm long and is black in colour. It lays its eggs on the soil surface above a corpse and the larvae hatch within 24 hours and then burrow down through soil for more than 2 metres to reach the corpse. They prefer to feed on lean, rather than fatty, tissue and then pupate and the adults emerge within 25 days. The fly can repeat its life cycle underground through many generations on the same corpse.

Leptocera (Figure 6.26) is a genus of flies that belongs to the family Sphaeroceridae, known as the "lesser corpse flies". They are small (1-5 mm

Figure 6.25   An adult Phorid fly, also known as a "scuttle fly". The coffin fly, *Conicera tibialis,* belongs to this group of flies.
U.S. Department of Agriculture, Public domain, via Wikimedia Commons

Figure 6.26   An adult male *Leptocera* species
Janet   Graham,   CC   BY   2.0   <https://creativecommons.org/licenses/by/2.0>,   via Wikimedia Commons

long) and dull-coloured. Their larvae feed on the microbes that live on corpses and dung.

Triphleba hyalinata, (Figure 6.27) like *Conicera tibialis,* is a member of the Phoridae family. As well as colonising buried corpses, it's also found in caves as well as on faeces. It seems to have a preference for corpses buried in wooden rather than metal coffins.

Figure 6.27 *Triphleba* species
Image courtesy of Derek S. Sikes, University of Alaska Museum of the North, CC BY 4.0

6.3.2 Now for the Late-Comers

Later colonisers (after 100 days of burial) include *Calliphora vicina, Megaselia rufipes* and *Omalium rivulare*. *Calliphora* species have been described earlier in this chapter (Section 6.2.1). *Megaselia rufipes,* like *Conicera tibialis,* is a type of scuttle fly and is also known as the "coffin fly" (Figure 6.28). It runs in a very erratic manner, frequently changing directions. Its larvae are white and, as well as being found on corpses (both buried and left above ground), are also frequently present on a wide range of decaying matter and faeces. The larvae are frequently found on the outer surfaces of the corpse or in its clothing.

Omalium rivulare is a rove beetle and these have been described above (Section 6.2.2). It is 3.5 - 4.00 mm long and has a brown colour (Figure 6.29).

6.3.3 And Last, But Not Least

The last wave of organisms to arrive, after about 200 days, include mites, *Fannia scalaris, Fannia manicata* and *Philonthus* species.

Mites (Figure 6.30) aren't insects but are a type of Arachnid i.e. an invertebrate with 8 jointed legs, no wings or antennae and a body divided into 2 parts. They are very small, often microscopic, and range in size from 0.1 to 6 mm. There are more than 48,000 species of mites and they are difficult to identify which is why many studies of corpse decomposition haven't named

(a)   (b)

Figure 6.28 Images of *Megaselia*. (a) An adult, male *Megaselia rufipes* (b) pupa of *Megaselia*

(a) Image courtesy of Steve Kerr and the Global Biodiversity Information Facility. CC BY 4.0.

(b) Alcaine-Colet A, Wotton KR, Jimenez-Guri E. Rearing the scuttle fly *Megaselia scalaris* (Diptera: Phoridae) on industrial compounds: implications on size and. *Peerj* 2015 ;3:e1085. DOI: https://doi.org/10.7717/peerj.1085. Creative Commons CC-BY 4.0

Figure 6.29 *Omalium rivulare*
Image courtesy of Donald Hobern and the Global Biodiversity Information Facility. CC BY 4.0.

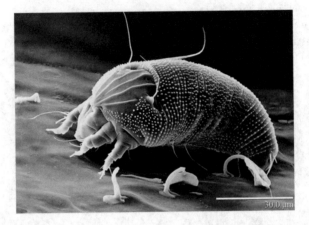

Figure 6.30 A mite as seen through an electron microscope. Note the scale which shows that the mite is only about 0.09 mm.
Photo by Eric Erbe; digital colorization by Chris Pooley (USDA, ARS, EMU).[2], Public domain, via Wikimedia Commons

which species are present. Collectively, they can use a very wide range of food sources including microbes, insects, worms and other mites. Some can also feed on dry skin and this enables them to become more dominant in the later stages of corpse decomposition.

It's beyond the scope of this book to describe all of the many species of mites (more than 75) that have been found on human corpses at various stages of decomposition. However, genera that have been found in the skeletonization stage include the following *Holostapsis, Hypoaspis, Laelaps, Gamasus, Uropoda, Acarus, Tyrophagus* and *Hoplophora* (Figure 6.31)

Fannia scalaris is a Muscid fly (described in Section 6.2.2) and is known as the "latrine fly" because it's often found in human faeces, as well as on corpses. It's 6-7 mm long and black in colour although its abdomen and thorax are silvery-grey (Figure 6.32). It has three longitudinal stripes on its thorax.

Its larvae are white or cream-coloured and are small (< 6 mm long) and flattened. It's life cycle can take 15 – 30 days depending on temperature and other factors and is summarised in Figure 6.33.

Fannia manicata, also known as the "lesser house fly" is similar in appearance to *Fannia scalaris* and has a similar life cycle.

Philonthus species (Figure 6.34) are a type of rove beetle which have been described previously (Section 6.2.2). There are about 380 species in the genus and they range in size from 5-13 mm. They are associated with decaying matter and are found in dung and compost as well as on corpses.

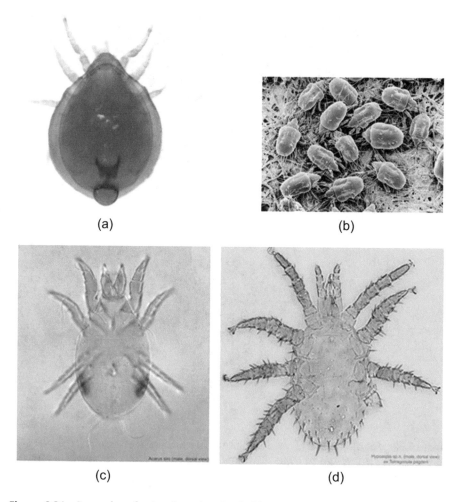

(a)

(b)

(c)

(d)

Figure 6.31 Examples of mites found on buried human corpses.
(a) *Uropoda* species – about 4 mm long
S.E. Thorpe, Public domain, via Wikimedia Commons
(b) *Tyrophagus* species (X400)
United States Department of Agriculture, Public domain, via Wikimedia Commons
(c) *Acarus* species
Photo by Pavel Klimov, Bee Mite ID (idtools.org/id/mites/beemites), Public domain, via Wikimedia Commons
(d) *Hypoaspis* species
Photo by Pavel Klimov, Bee Mite ID (idtools.org/id/mites/beemites), Public domain, via Wikimedia Commons

(a)

(b)

Figure 6.32 *Fannia scalaris* (a) an adult insect (b) drawing of an adult
(a) Permission is granted to copy, distribute and/or modify this document under the terms of the GNU Free Documentation License, Version 1.2 or any later version published by the Free Software Foundation; with no Invariant Sections, no Front-Cover Texts, and no Back-Cover Texts.
(b) Unknown author, Public domain, via Wikimedia Commons

So, now we've seen how both microbes and insects utilise the valuable nutrients locked away in our dead bodies. In death, we've certainly contributed to the proliferation of both of these huge kingdoms of life. But what other effects does our corpse have on the environment? This question will be considered in the next chapter.

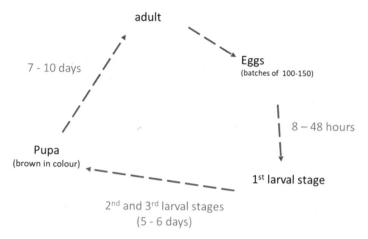

Figure 6.33 Life cycle of *Fannia scalaris*

Figure 6.34 An adult *Philanthus* species
Image courtesy of Alexis and the Global Biodiversity Information Facility. CC BY 4.0.

6.4 Want to Know More?

Lots of information about insects in general as well as descriptions of particular species. Amateur Entomologists' Society
https://www.amentsoc.org/insects/

General information about insects as well as about the Smithsonian's National Museum of Natural History's insect collection
https://www.si.edu/spotlight/buginfo

Insects: the most successful lifeform on the planet.
https://www.earthlife.net/insects/six01.html

Morphology and identification of fly eggs: application in forensic entomology.
Sanit S et al. *Tropical Biomedicine* 2013 Jun;30(2):325-37.
http://www.msptm.org/files/325_-_337_Sukontason_K.pdf

A brief review of forensically important flesh flies (Diptera: Sarcophagidae).
Ren L *et al. Forensic Science Research* 2018 Mar 22;3(1):16-26. doi:
https://doi.org/10.1080/20961790.2018.1432099.
https://pubmed.ncbi.nlm.nih.gov/30483648/

The use of insects in forensic investigations: An overview on the scope of
forensic entomology. Joseph I *et al., Journal of Forensic Dental Science* 2011
Jul;3(2):89-91. doi: https://doi.org/10.4103/0975-1475.92154
https://www.ncbi.nlm.nih.gov/pmc/articles/PMC3296382/

Pigs vs people: the use of pigs as analogues for humans in forensic entomology and taphonomy research. Matuszewski S et al., *International Journal of
LegalMedicine*2020Mar;134(2):793-810.doi:https://doi.org/10.1007/s00414-019-
02074-5. Epub 2019 Jun 17
https://pubmed.ncbi.nlm.nih.gov/31209558/

A biological and procedural review of forensically significant *Dermestes* species
(Coleoptera: Dermestidae). Magni PA et al. *Journal of Medical Entomology*,
2015; 52(5):755-769. https://doi.org/10.1093/jme/tjv106

Forensic Entomology: Effective role of insects in death investigation, 2020
https://forensicyard.com/forensic-entomology/

Forensic Entomology. EDInformatics.com
https://www.edinformatics.com/forensic/forensic_entomology.htm

Forensic Entomology: The use of insects in criminal investigations.
Science Monk
https://sciencemonk.com/forensic-entomology-the-use-of-insects-
in-criminal-investigations/

Forensic entomology: arthropods at the crime scene
https://allyouneedisbiology.wordpress.com/tag/worm-on-corpses/

Characterizing forensically important insect and microbial community colo-
nization patterns in buried remains. Iancu L. *et al. Scientific Reports* 2018;
volume 8, 15513
https://www.nature.com/articles/s41598-018-33794-0.pdf

Time flies - age grading of adult flies for the estimation of the post-mortem interval
Amendt J *et al., Diagnostics (Basel).* 2021 Feb; 11(2): 152. doi: https://doi.org/10.3390/diagnostics11020152
PMCID: PMC7909779
https://www.ncbi.nlm.nih.gov/pmc/articles/PMC7909779/

Forensic entomology: applications and limitations. Amendt J *et al. Forensic Science, Medicine and Pathology* 2011 Dec;7(4):379-92. doi: https://doi.org/10.1007/s12024-010-9209-2.

Effectiveness of chronic wound debridement with the use of larvae of *Lucilia Sericata*. Bazaliński D et al., *Journal of Clinical Medicine* 2019 Nov 2;8(11):1845. doi: https://doi.org/10.3390/jcm8111845. PMID: 31684038
https://www.ncbi.nlm.nih.gov/pmc/articles/PMC6912827/

A systematic review of maggot debridement therapy for chronically infected wounds and ulcers. Sun X *et al., International Journal of Infectious Diseases* 2014 Aug;25:32-7. doi: https://doi.org/10.1016/j.ijid.2014.03.1397. Epub 2014 May 16. PMID: 24841930
https://www.ijidonline.com/article/S1201-9712(14)01494-5/fulltext

Changing attitudes toward maggot debridement therapy in wound treatment: a review and discussion. King C. *Journal of Wound Care* 2020 Feb 1;29(Sup2c):S28-S34. doi: https://doi.org/10.12968/jowc.2020.29.Sup2c.S28.

Degradation of extracellular matrix components by defined proteinases from the greenbottle larva *Lucilia sericata* used for the clinical debridement of non-healing wounds. Chambers L *et al. British Journal of Dermatology* 2003;148(1):14-23.

Larval therapy for leg ulcers (VenUS II): randomized controlled trial. Dumville JC *et al. British Medical Journal 2009*; 338, 1047–1054.

Entomofauna of buried bodies in northern France. Bourel B *et al. International Journal of Legal Medicine* 2004 Aug;118(4):215-20. doi: https://doi.org/10.1007/s00414-004-0449-0. Epub 2004 Apr 28.

Determination of post-burial interval using entomology: A review. Singh R, Sharma S, Sharma A. *Journal of Forensic and Legal Medicine* 2016 Aug;42:37-40. doi: https://doi.org/10.1016/j.jflm.2016.05.004. Epub 2016 May 12

Forensic entomology investigations from Doctor Marcel Leclercq (1924–2008): a review of cases from 1969 to 2005. Dekeirsschieter J. *et al., Journal of Medical Entomology*, 2013; 50(5) : 935-954

7

And What About the Rest of the Big, Wide World? Corpse Decomposition and the Environment

So far this book has been very much human-focussed – all we've been bothered about is what happens to our dead body. Let's now go beyond this and start looking at the wider environment. We've seen in Chapters 5 and 6 that a variety of microbes and insects make full use of the nutrients present in a human corpse. Now that you're an ecologist (as well as a biochemist, microbiologist and entomologist) you're going to want to know what effect all this has on the environment in which the corpse has been left. That corpse decay does have a significant environmental impact can best be appreciated if we get away from human corpses for a moment. It's been estimated that natural death results in 5,000 kilograms of mammalian tissue being deposited on each square kilometre of soil per year – that represents an awful lot of nutrients for hungry microbes and insects. Now, back to humans. Each year approximately 58 million people die and the majority of these are buried, or their cremated remains are added to the environment – that's certainly going to have an impact. The approximate quantities of the major nutrients present in this number of dead humans is shown in Figure 7.1

So, from this figure, you can appreciate that such enormous quantities of nutrients (for example 742 million kilogrammes of protein) will be able to feed huge numbers of microbes, insects and other creatures.

Depositing a corpse, with its large and concentrated nutrient content, into the environment is certainly going to affect that environment in a number of ways. We've already seen that both the microbial and insect communities living on the corpse go through a series of changes in their composition and these will have consequences for the immediate environment surrounding the corpse. This is because both the microbial and insect communities are not

© The Author(s), under exclusive license to Springer Nature Switzerland AG 2022
M. Wilson, *Life After Death: What Happens to Your Body After You Die?*,
Springer Praxis Books, https://doi.org/10.1007/978-3-030-83036-6_7

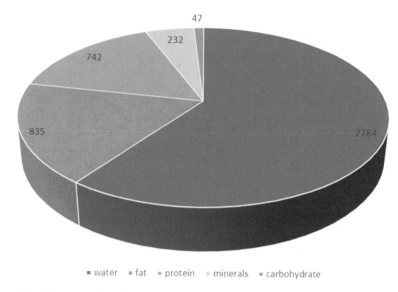

Figure 7.1 The quantities (in millions of kilograms) of major nutrients present in the corpses of the 58 million people who die each year

physically confined to the corpse itself but will spill over into the surroundings. There they'll have an impact on the organisms that live there as well as on its chemical composition and, hence, its ability to support life. But there'll also be more widespread consequences, because the insects that visit the corpse, as well as the adults that emerge from the eggs that are laid in it, will spread into the wider environment where they will have a variety of effects. These will include damage to plants and animals, as well as important beneficial outcomes such as bringing about pollination and acting as a food source for a variety of animals.

So, in order to appreciate the environmental impact of a corpse we have to consider a number of components: the short and long-term consequences as well as the local and more distant effects (Figure 7.2).

7.1 What Are the Ways in Which a Corpse Affects the Environment?

For convenience, we can classify these impacts as sensual, chemical and biological. However, this is somewhat artificial and there's going to be some overlap between them (Figure 7.3). These impacts will change the environment in which the corpse has been placed in a number of ways. For example, the

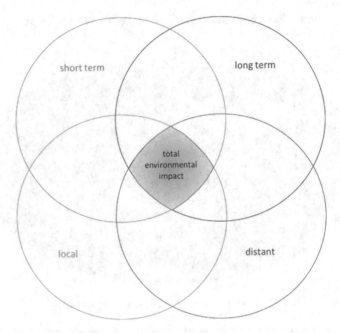

Figure 7.2 The various components of the environmental impact of a corpse

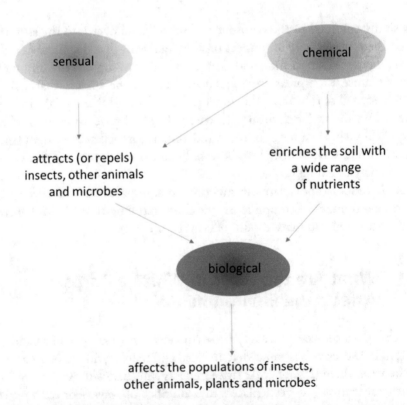

Figure 7.3 The various types of impact that a corpse has on its environment

chemical composition of the surrounding soil will be altered, the populations of particular creatures will be affected, the growth of one or more plants will be stimulated (or inhibited), the behaviour and/or nutritional status of animals will be changed etc.

7.1.1 Let's Start With the Sensual

A corpse left to rot above ground has an immediate visual impact – for humans this is invariably considered to be unpleasant. However, for many animals a corpse is a welcome sight – it signals the presence of food. We've seen in Chapter 4 that, for most of us, a rotting corpse isn't a pleasant site, particularly during the early stages. However, once it's reached the skeletonisation stage, most of us can probably just about cope with the site of a few bones. Depending on how well our eyes function, the visual impact is likely to be local and a corpse won't be visible to us from a distance. Some animals, however, have greater visual acuity and use sight as one of their main means of detecting food sources. For example, flies are known to use their eye sight to detect corpses. Birds that feed on corpses need to be able to spot them from long distances. Vultures have been reported to be able to spot a 3 foot long object from a distance of 6 kilometres and so will be able to detect a corpse from far away. This means that the visual impact of a corpse can have both short- and long-term environmental consequences both locally and over large distances.

When it comes to smell, some animals can detect a corpse by its odour from many kilometers away. For humans, the smell of a corpse during the first four stages is extremely unpleasant, but we're generally unable to detect any smell once it's reached the skeletonisation stage. A large number of odorous compounds are produced by a corpse and these have been described in Section 4.4. The odours we find most offensive include those of sulphur-containing compounds (hydrogen sulphide, methanethiol), various amines (putrescine and cadaverine) and certain organic acids (butyric and propionic acids). While we can detect such compounds over a limited distance (a few yards at most), other animals are far more sensitive. Suitably-trained dogs, for example, can detect a corpse at a distance of more than 3 Kilometers. Some insects (blow flies) have been reported to be able to detect a corpse 16 Kilometers away. Vultures can use their sense of smell to locate corpses that are deep inside forests and so can't be seen from above. Corpse odour, therefore, can have a wide-ranging impact on the environment as it will attract a variety of animals from near and far to this important nutrient source. However, this will be limited to the short-term because odorous compounds aren't produced after the first 4 stages of decomposition. Microbes can also be considered to have a sense of smell, although this is very primitive. They have receptors on their surface that can detect a wide range of compounds and respond by moving toward them. This will be effective in the long-term as well as in the short-term.

Let's not even think about taste, far too awful to contemplate. And, although a rock group called The Move in 1967 sang that they "can hear the grass grow", it's doubtful that even they would claim to be able to hear a corpse decompose. The rattling of bones being moved by the wind? Maybe. Perhaps Peter, Paul and Mary would be able to, as in 1962 they claimed to be able to "hear a whistle blow 500 miles" away.

7.1.2 Now for the Chemical

Our interest in chemistry in this book so far has been focussed on the ways in which large molecules can be broken down by enzymes to provide nutrients for microbes and insects. However, many of these compounds (e.g. amino acids, sugars etc.) won't get consumed by living creatures immediately but will seep into the ground in body fluids and so alter the chemical composition of the surrounding soil. Furthermore, many of the minerals present in body fluids will likewise drain away into the soil. This will all happen in the short term, during the first four stages of decomposition. It's been shown that 0.4–0.6 litres of fluid per kilogram of body weight leach into the soil during the course of decomposition. The composition of this fluid is summarised in Figure 7.4.

During the skeletonization stage, and subsequently, the bones and teeth will gradually decay and their mineral constituents (mainly calcium and phosphate ions) will be absorbed into the soil where they serve as valuable nutrients for plants. So, there are short- and long-term impacts, although their effects will mainly be local – except when pieces of the corpse (or bones) are removed by scavengers (Box 7.1) and transported away from the corpse.

Then, of course, there are the volatile compounds produced during decay that the wind can carry far from the corpse. As we discussed earlier, these can act as attractants for insects, birds and other animals so enabling them to find the corpse.

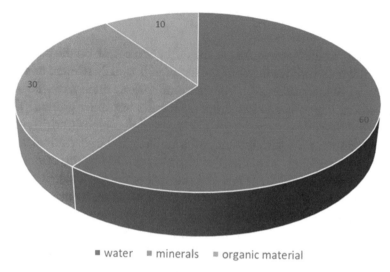

water   minerals   organic material

Figure 7.4 Composition of the fluid that leaches into the soil during the decomposition of a human corpse. Minerals are in the form of ions containing nitrogen, phosphorus, chloride, bicarbonate, calcium, sodium, titanium, chromium, iron and other metals

Box 7.1. Environmental impact of scavengers

A predator is an organism that lives by preying on other live organisms while a scavenger is one that feeds on dead or decaying organisms that it hasn't killed itself. There are two types of scavengers – those that feed only on dead organisms (obligate scavengers) and those that are mainly predators but will also eat dead organisms if these are available (facultative scavengers). Obligate scavengers include vultures, condors, burying beetles, blowflies and wasps. Examples of facultative scavengers include hyenas, crows, lions, leopards and wolves.

Any corpse left out in the open is a potential source of nutrients for insects, microbes and scavengers. Although scavengers are rarely involved in the disposal of the corpses of humans, this isn't the situation when it comes to other dead animals. Scavengers are estimated to consume between 35 and 75% of the cadavers left on the land and this can take place very quickly. On the grass plains of East Africa, for example, scavengers such as vultures, hyenas and jackals can reduce a corpse to scattered bones in just a few hours (an example of this can be seen in Appendix III). This means that very little decomposition would have taken place and so few nutrients would have entered the soil.

In temperate climates during the winter, up to 100% of corpses may be eaten by scavengers. This is because insects and microbes are much less active at low temperatures leaving scavengers much more time to find the corpses before they have decomposed. This would result in a substantial reduction in the quantity of nutrients entering the soil during the winter. However, in the case of the corpses of large animals, often the bones, hair and skin would be left behind to decompose which would result in some enhancement of soil fertility.

7.1.2.1   Local Effects

A number of studies have looked at the effect of a corpse on the composition of the soil with which it's in contact – such soil is usually referred to as "grave soil". As in many studies of corpse decomposition, these have, for ease of investigation, largely been carried out on corpses left above ground. The soil is affected differently during the various stages of decomposition. The chemical composition of the surrounding soil is unlikely to be affected by the presence of a corpse until it reaches the bloat stage. During this stage maggot activity produces lots of ammonia and this makes the soil alkaline which can kill neighbouring plants. The soil will also become richer in ammonium ions which are important nutrients for plants. During the active decay stage, purge fluids that escape from the mouth, nostrils and anus will seep into the soil. These fluids contain a wide variety of inorganic and organic materials and so will enrich the soil to some extent. The rupturing of the skin that occurs during this stage releases much greater quantities of body fluids that are absorbed by the soil. At this stage, in the case of bodies that have been left above ground, the environmental impact of the decaying corpse becomes apparent as a distinctive "halo" of dead plant material surrounding it. This is known as a "cadaver decomposition island" (CDI – Figure 7.5 – a picture of this can be seen in Appendix III) and is the result of plant death (due, in part, to the increased alkalinity) and the destruction caused by maggots migrating away from the corpse in order to pupate. The average human corpse creates a CDI with an area of about 3 square metres. But don't be dismayed, although this looks awful, this short-term destruction masks a dramatic increase in soil fertility (discussed below) and is only a temporary effect. The richly-fertilised soil will later support the growth of many plants.

Notable changes are apparent in the composition of soil beneath a corpse during the various stages of decomposition (Figure 7.6).

From Figure 7.6 it can be seen that the total organic carbon, total nitrogen and ammonium ion content of the soil started to increase during the bloat-active transition and peaked during the early advanced stage. This corresponds to the leakage of body fluids into the soil during these stages. The pH of the soil became alkaline during the bloat-active transition and at the mid-advanced stages. Other studies have shown a more dramatic change in soil pH and an increase of 2.0 pH units is frequently seen. Because the pH scale is logarithmic, this represents a one hundred-fold increase in the pH of the soil environment.

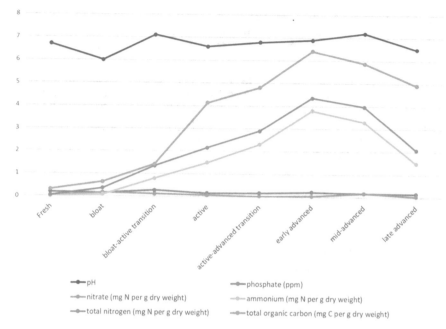

Figure 7.6   Trends in the concentrations of various chemicals in the soil layer (0-3 cm) immediately beneath four human corpses left in the open air

The presence of a corpse also affects soil composition in the long term. The results of a long-term study of the soil beneath corpses are shown in Table 7.1. This study analysed the concentration of various chemicals in the 5 cm layer of soil beneath corpses for a period of nearly 5 years. The chemicals chosen were those that are important plant and microbial nutrients such as nitrate, ammonium, phosphate, sodium, potassium, magnesium and calcium. They also included chemicals that would be of particular importance to soil microbes such as dissolved organic nitrogen compounds and dissolved organic carbon compounds. As can be seen from the table, in the case of each chemical, a peak concentration was reached at a particular PMI (when it was released from the corpse) and this was followed by a decrease. Sometimes this occurred at an early PMI (e.g. 35 days in the case of sodium) while for others this happened much later (e.g. 350 days for nitrate). In some cases, the concentration of the chemical in the soil was considerably higher than it was before the corpse was deposited there. For example, the concentrations of nitrate and ammonium ions were around 100 times greater. After varying periods of time, the concentration of the chemical often returned to what it was before the corpse was deposited – this was usually between 650 and 800 days. However, in some cases (phosphate, dissolved organic nitrogen and dissolved organic

Table 7.1 Changes in the concentrations of various chemicals in the 5 cm layer of soil beneath a corpse over a period of approximately 5 years

Chemical	General pattern during the course of the study	Number of times that the peak concentration was greater than that of the pre-corpse level	Approximate time at which the concentration returned to the pre-corpse level (days)
Sodium	Reached a peak at about 35 days then gradually decreased	9	767
Potassium	Reached a peak at about 150 days then gradually decreased	6	750
Magnesium	Reached a peak at about 120 days then gradually decreased	2	700
Calcium	Reached A Peak At About 300 Days Then Gradually Decreased	3	700
Nitrate	Reached a peak at about 350 days then gradually decreased	100	684
Ammonium	Reached a peak at about 177 days then gradually decreased	90	659
Phosphate	Reached a peak at about 132 days then gradually decreased	70	>1752
Dissolved organic nitrogen	Reached a peak at about 132 days then gradually decreased	30	>1752
Dissolved organic carbon	Reached a peak at about 132 days then gradually decreased	30	>1752

carbon) the concentrations remained higher than the pre-corpse values even after 1752 days. Other studies have also reported that high levels of nutrients persist in the soil for 5 years or more. The gradual decrease in concentration following the initial peak is due to a number of factors. First of all, the chemical will gradually diffuse away from the corpse and, secondly, it will be used up by the microbes and/or plants that are present.

Analysis of the corpse soil at a depth of 10-15 cm showed that the concentrations of all chemicals were increased compared to the pre-corpse deposition values. In the cases of nitrate and ammonium, high levels persisted for as long as 715 days.

To summarise, a decomposing corpse has a huge impact on the chemical composition of the surrounding soil both in the short-term and for a number of years later. The soil becomes enriched with respect to a wide range of chemicals that are important plant nutrients and that can also be used by many soil microbes.

7.1.2.2   Distant Effects

The diffusion of chemicals away from a single corpse in all directions means that within a few metres they're unlikely to exert any significant effect and the chemical composition of the soil is unlikely to be affected. However, what happens when there are large numbers of corpses in a relatively small area such as in a cemetery? Might the substances released from this high concentration of decomposing corpses have a significant effect on the chemical composition of the soil? In the UK a typical cemetery will have approximately 2,000 burials per hectare. This means that the graves represent less than 1% of the total mass of the cemetery and about 7% of the site volume down to a depth of 1.8 metres. Not surprisingly, many studies have found higher concentrations of a variety of corpse-derived chemicals in the soils of cemeteries. In addition to such chemicals, substances that originated from materials associated with burial practices such as coffin materials and embalming procedures, were also found. Concern, therefore, has been expressed about the possibility that these chemicals could find their way into ground water and so contaminate water supplies. Consequently, a number of studies have been carried out to see if this happens. It's difficult to generalise from the results of these studies because the extent of any contamination of water supplies is affected by many variables such as the density of graves in the cemetery, the geology of the cemetery site, the distance of the cemetery from the ground water or water course and the prevailing climate (particularly with regard to rainfall). Major findings from such studies are summarised in Table 7.2.

In a 2012 review of data obtained from studies carried out worldwide, it was concluded that cemeteries did adversely affect the quality of underground waters and this occurred to the greatest extent in cemeteries located in warm

Table 7.2 Summary of main findings from studies of the contamination of groundwaters near cemeteries

Country	Groundwater contained increased concentrations of:
United Kingdom	sulphates, chlorides, sodium, carbolic acid, zinc, ammonium, copper, manganese, zinc and iron
Canada	nitrate, nitrite and phosphate
Brazil	chloride, nitrate, bicarbonate, sodium, calcium, iron, aluminium, lead and zinc
Australia	nitrate, phosphate, ammonium, nitrate, orthophosphate, chloride, bicarbonate, iron, sodium, magnesium, zinc and potassium
Poland	phosphate, iron, manganese, copper, zinc and aluminium
Portugal	heavy metals, including lead and zinc
Netherlands	chlorides, sulphates, bicarbonates

and humid climatic regions such as the Republic of South Africa and Brazil. Most researchers regarded nitrogen- and phosphorus-containing ions from cemeteries as posing the greatest threat to water quality.

7.1.3   The Biological Impact

Having seen that chemicals leach out from the decomposing corpse and are absorbed by the soil, it should come as no surprise to learn that this has a profound effect on the organisms that live there. In Section 5.6.3 we learnt that the creatures that inhabit the soil are classified on the basis of their size into three groups - microbiota (<0.2 mm), mesobiota (0.2-10.0 mm) and macrobiota (>10.0 mm). All three groups will be affected by changes in the chemical composition of the soil that have been brought about by the decomposing corpse.

7.1.3.1   What Happens to the Soil Microbiota?

First of all, let's take a look at the microbiota. Before we do that, it's important to appreciate that the microbiota of the soil is even more complex and diverse than that of a live human which we discussed in Chapter 3 - there can be as many as 8 million different bacterial species in every gram of soil. Because of this, detailed analysis of the soil microbiota at the genus level would be difficult to describe in a way that's suitable for this book. Also, this would mean introducing a long catalogue of new genera that are very different from those we've met so far – and I don't want to confuse you even more. Consequently, we're generally going to look at the microbiota at only the level of the phyla rather than the families or genera.

Figure 7.7 shows the results of an investigation into the effects of corpse decomposition on the microbial communities present in the 3 cm layer of soil immediately underneath the corpse. From this you can see that during the fresh phase, the soil microbiota remained similar to that which was present before the corpses were placed on the soil – the dominant phyla being Acidobacteria and Proteobacteria. The proportion of Firmicutes then gradually increased and peaked at the early advanced stage, then decreased. In contrast, the proportions of Acidobacteria and Verrucomicrobia gradually decreased with increasing PMI. The proportion of Proteobacteria remained fairly constant throughout the various stages of decay. Interestingly, during the active decay stage the proportions of certain human-associated genera peaked in abundance. These included *Staphylococcus* and *Enterococcus* and are evidence of the movement of bacteria from the corpse into the environment. As both of these genera belong to the Firmicutes phylum, they will have contributed to the observed increase in the proportion of this phylum during that decomposition stage. During the early advanced decay stage there were increased proportions of genera from the human gut – *Lactobacillus* (from the

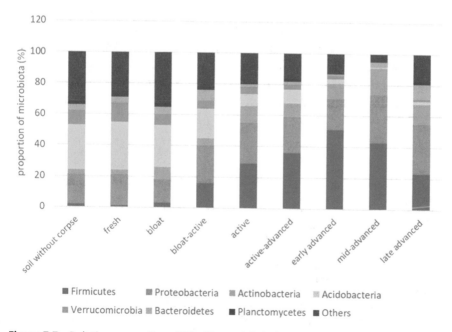

Figure 7.7   Relative proportions (%) of bacterial phyla that are present in the soil layer (0 - 3 cm) immediately beneath four human corpses left to decompose in the open air

Firmicutes phylum), *Phascolarctobacterium* (from the Firmicutes phylum) and *Eggerthella* (from the Actinobacteria phylum). Again, this will have contributed to the observed increases in the proportions of these phyla.

Studies have shown that the soil microbiota remains disturbed by the presence of a corpse for as long as 2 years and generally show an increased proportion of Firmicutes, due to the movement of human-associated bacteria into the soil, and a decreased proportion of Acidobacteria.

A concern with regard to corpse-derived microbes is the possibility that they not only enter the soil but might also find their way into water supplies. Why should this be a problem? The answer is that some members of the human microbiota have the potential to cause disease if they are ingested in drinking water; the elderly, the very young and those who are immunocompromised in some way are particularly vulnerable in this respect. Examples of such microbes include *Staphylococcus aureus*, *Streptococcus pyogenes*, *Streptococcus pneumoniae*, *Neisseria meningitidis*, *Clostridium* species, *Bacteroides* species and *Enterococcus* species. Also, if a person has died from an infectious disease, then the corpse may contain high concentrations of disease-causing bacteria, viruses, fungi or protozoa. There is certainly evidence that diseases can be transmitted in this way, particularly in the case of the latter scenario i.e. the corpses of people who have died of an infectious disease. For example, people have caught typhoid fever by drinking water contaminated with *Salmonella typhi* which had come from nearby cemeteries in which typhoid victims had been buried. This occurred in Paris (in 1879 and in 1954) and in Berlin (during the period 1963 - 67). The monitoring of groundwater near cemeteries has revealed the presence of high concentrations of microbes that are capable of causing disease and the results of some of these studies are summarised in Table 7.3.

Table 7.3 Examples of studies that have detected microbes in groundwater near cemeteries

Country	Microbes detected in groundwater
England	*Enterococcus faecalis*, *Bacillus cereus*, *Clostridium perfringens*, *Staphylococcus aureus*
Poland	*Bacillus cereus*, *Staphylococcus aureus*, *Staphylococcus* species, *Clostridium perfringens*, *Enterococcus faecalis*, *Escherichia coli*
Brazil	*Pseudomonas* species, *Bacillus* species, *Salmonella* species, *Escherichia coli*, *Enterococcus* species, *Clostridium* species
South Africa	*Enterococcus faecalis*, *Staphylococcus aureus*, *Escherichia coli*
Australia	*Enterococcus faecalis*, *Pseudomonas aeruginosa*, *Escherichia coli*

In general, microbial contamination of groundwater near cemeteries occurs more frequently, and at higher levels, in countries with warm, moist climates. The World Health Organisation recommends the following measures to reduce the possibility of microbial contamination of water supplies:

- corpses should not be buried within 250 metres of any well, borehole or spring from which a potable water supply is drawn.
- the base of any burial pit should be at least one metre above solid rock.
- the base of all burial pits must be at least one metre above the highest natural water table.

7.1.3.2   Now for the Nematodes

Not much is known about the effects that a decomposing corpse has on the macrobiota of soil. However, the populations of nematodes (a major group belonging to the mesobiota) are known to be dramatically altered by the presence of a corpse within a few days. Nematodes, also known as roundworms or threadworms, are thin and elongated with a tubular body that has no distinct head or tail (Figure 7.8).

Those that live in the soil can be broadly classified on the basis of their food preferences into bacterial feeders, fungal feeders, plant feeders and those that are parasites for a wide range of creatures. The presence of a corpse alters the relative proportions of these various types of nematodes in a soil, and within

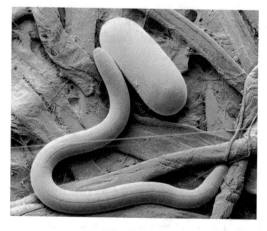

Figure 7.8 Scanning electron micrograph of a typical nematode and one of its eggs (x1000)
Agricultural Research Service, Public domain, via Wikimedia Commons

Figure 7.9 *Caenorhabditis elegans*, a nematode that feeds mainly on bacteria
Image courtesy of the Okinawa Institute of Science and Technology Graduate University,
Japan. This work is licensed under a Creative Commons Attribution 2.0 Generic License

a few days the soil becomes dominated by species that feed on bacteria. Examples of these include species belonging to the genera *Acrobeloides, Bursilla, Cephalobus, Cruznema, Rhabditis, Caenorhabditis* and *Panagrolaimus.* They are voracious consumers of bacteria and can eat about 5×10^6 bacteria per day. One of the most extensively studied species of these bacteria-feeders is *Caenorhabditis elegans* (Figure 7.9).

Nematodes improve soil fertility by converting the organic nitrogen compounds that are present in the microbes and plants they eat into ammonium ions that can be used as a source of nitrogen by plants.

7.1.3.3   The Micro-Arthropods Are Also Affected

The presence of a human corpse also affects another major group of the mesobiota – the micro-arthropods. These are small invertebrates such as springtails and mites. They are important in maintaining soil fertility because they consume dead leaves, as well as microbes, and convert them into smaller compounds and into minerals that can be used by plants. In one study, these were found to be 15 - 17 times more abundant in soils collected from beneath human corpses. In particular, there were increased proportions of members of microbe-consuming species belonging to families such as Isotomidae (Figure 7.10a), Uropodidae (Figure 7.10b) and Acaridae (Figure 7.10c).

7.1.3.4 What About Insects?

We've seen in Chapter 6 that a human corpse is an excellent supply of nutrients for a large variety of insects during the various stages of its decomposition. The decay of a corpse, therefore, will result in the production of very large numbers of insects. We've also seen in Chapter 6 that the insects themselves, especially their eggs and larvae, act as food sources for other insects that can't directly feed on the corpse itself – examples include Rove beetles, Clown beetles, *Hydrotaea* flies and Checkered beetles. While we know that on the corpse itself there are huge increases in the population of certain insects at particular stages in its decomposition, what are the wider effects? Are there changes in the insect populations in the environment beyond the corpse? If so, are they long-term? Unfortunately, studies that can answer these questions appear not to have been carried out so far.

7.1.3.5 Are Other Animals Affected?

Corpses are, of course, a source of food for scavengers. Unless we're a murder victim, or die by accident in some remote region, then we're unlikely to end up as food for scavengers. However, in some countries, followers of the Buddhist religion prefer what is known as a "sky burial" to being cremated or buried in the ground. This involves leaving the corpse out in the open (often the top of a mountain – Figure 7.11) to be eaten by birds of prey, usually vultures. When all that remains are bones, these are often then baked in a bread and left out to be eaten by birds. This is practiced in Tibet, Mongolia, Bhutan and certain parts of India and China. It was also practiced by some Native American tribes.

Non-scavenging animals can also benefit from corpses. This is because the insects that feed on a buried corpse will eventually fly away and many of them will be eaten by a wide range of animals including birds, fish, mammals (shrews, hedgehogs, moles etc), amphibians and reptiles. In some parts of the world, insects are also an important source of food for humans (Figure 7.12). Nearly 2,100 different species of insects are eaten by people in 130 countries

(a)

(b)

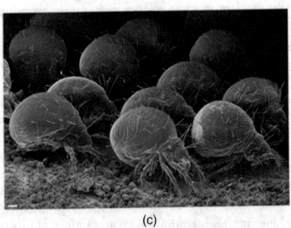

(c)

Figure 7.10 Examples of micro-arthropods whose populations increase in soil beneath a human corpse
(a) An example of a member of the Isotomidae (springtails), *Folsomia candida*, which is about 2.5 mm long
Faddeeva-Vakhrusheva, A. *et al*. Coping with living in the soil: the genome of the parthenogenetic springtail *Folsomia candida* . *BMC Genomics* 18, 493 (2017). https://doi.org/10.1186/s12864-017-3852-x

Figure 7.11   A buddha at a sky burial site in a monastery near Yushu, Tibet
(WT-en) Nomadsolicitor at English Wikivoyage, Public domain, via Wikimedia Commons

7.1.3.6   What About Plants?

Any corpse placed on top of plant-covered soil is going to have an immediate effect on the plants that are already there (Box 7.2). The corpse will crush any underlying plants and deprive them of sunlight and so many will die. Some, however, may be able to survive because they've produced seeds or they're equipped with underground structures that can grow again once the corpse has decomposed and sunlight can reach them. Plants, their seeds and their underground structures, can also be killed by the fluids that leak out from the corpse during its subsequent decomposition. All of this contributes to the

---

Figure 7.10 (continued) This article is distributed under the terms of the Creative Commons Attribution 4.0 International License
(b) *Oplitis pusaterii*, an example of a member of the Uropodidae (mites)
S.E. Thorpe, Public domain, via Wikimedia Commons
(c). Scanning electron micrograph of a group of *Archegozetes longisetosus* feeding on microbes. Scale bar: 100µm. This is a species of mite belonging to the Acaridae and is one of the most common type of soil mites.
The 20th anniversary of a model mite: a review of current knowledge about *Archegozetes longisetosus* (acari, oribatida). Eethoff M. *et al*. *Acarologia* 2013; 53: 353-368. This work is licensed under a creative commons attribution 4.0 international license

Figure 7.12 A variety of insects at a food stall in Krabi, Thailand
Image courtesy of Paul Arps via flickr. Attribution 2.0 Generic (CC BY 2.0)

characteristic cadaver decomposition island (Figure 7.5 in Appendix III) that becomes apparent during the later stages of corpse decomposition.

Box 7.2 Remote detection of corpses in forests

Forests cover about 31% of the Earth's land mass altogether and a large proportion of the temperate and tropical latitudes where most humans live. Many people disappear in forested areas and are presumed dead. Often ground-based searches for them are very difficult and remote sensing and detection of decomposing human bodies is very difficult because of the forest canopy. However, local environmental disturbances caused by the burial of bodies, or their decomposition on the surface, alters the local physical and chemical environment and, inevitably, the local vegetation. Such changes may be detectable remotely. For example, the huge increase in soil nitrogen concentration during corpse decomposition can increase the chlorophyll content of the leaves of plants in the vicinity. Such changes can be detected by a device known as a spectrometer which measures the wavelengths present in the light that is emitted by, or reflected from, an object – this is known as the "spectrum" of that object. Cadmium is an element that accumulates in humans during life – especially in those who smoke or live in urban areas with high levels of pollution. The element is usually present at low concentrations in soil but is increased near a decaying corpse. Plants readily take up cadmium and this changes their leaf spectrum.

An unmanned aerial vehicle, or drone (Figure), equipped with a spectrometer could be used to scan large areas of forest and detect those changes in the leaf spectra of plants that are suggestive of the presence of a human corpse. This would significantly aid the detection of human bodies in forested regions.

Box 7.2 (continued)

Figure. An example of an unmanned aerial vehicle
Jason Blackeye jeisblack, CC0, via Wikimedia Commons

So far, so bad. However, there's another side to all this. In the long term, the previously-described (Section 7.1.2) increase in the concentration of important plant nutrients (nitrate, phosphate, metal ions etc) in the soil beneath, or surrounding, a corpse results in a stimulation of plant growth and this effect can last for several years. Animal bones are known to increase plant growth and so have been used as a fertiliser for hundreds of years (Box 7.3).

Box 7.3 Bones as a fertiliser

In the early 19th century, studies by scientists such as Justus von Liebig in Germany and John Bennet Lawes in England established the importance of nitrate, phosphate and various metal ions in plant nutrition. Their work was the basis for the formulation of effective fertilisers and the growth of what is now a huge industry responsible for their production. Around this time it was discovered that bones contained large quantities of two important plant nutrients - calcium and phosphate. Consequently, crushed or burnt bones were spread onto fields in order to improve their fertility. It was soon recognised, however, that the precious nutrients in bone dissolved only very slowly in water so weren't readily available to plants. In the 1830s von Liebig discovered that bones could be dissolved in sulphuric acid and that the resulting products were soluble in water and so could be absorbed by the roots of plants. The success of this new fertiliser created a huge demand for bones. The bones of buffaloes were collected from the North American prairies (Figure a), camel bones from the Egyptian desserts and human bones from the battlefields of Europe (Waterloo, Crimea and Leipzig).

Box 7.3 (continued)

Figure (a) Buffalo bones gathered from the Prairie for shipment, at Gull Lake, Alberta, Canada. 1884
Image from Major Matthews collection, City of Vancouver Archives. Public Domain

This prompted von Liebig to make the complaint that: "Great Britain was like a ghoul, searching the continents for bones to feed its agriculture...."

War and Commerce.—It is estimated that more than a million of bushels of human and inhuman bones were imported last year from the continent of Europe, into the port of Hull. The neighbourhood of Leipsic, Austerlitz, Waterloo, and of all the places, where, during the late bloody war, the principal battles were fought, have been swept alike of the bones of the hero, and of the horse which he rode. Thus collected from every quarter, they have been shipped to Hull, and thence forwarded to the Yorkshire bone-grinders, who have erected steam-engines and powerful machinery, for the purpose of reducing them to a granulary state. In this condition they are sent chiefly to Doncaster, one of the largest agricultural markets in that part of the country, and are there sold to the farmers to manure their lands. The oily part gradually evolving as the bone calcines, makes a more substantial manure than almost any other substance, and this is particularly the case with human bones. It is now ascertained beyond a doubt, by actual experiment upon an extensive scale, that a dead soldier is a most valuable article of commerce; and, for ought known to the contrary, the good farmers of Yorkshire are, in a great measure, indebted to the bones of their children for their daily bread. It is certainly a singular fact, that Great Britain should have sent out such multitudes of soldiers to fight the battles of this country upon the continent of Europe, and should then import their bones as an article of commerce to fatten her soil!

Box 7.3 (continued)

Reproduced in Figure b is part of an article from a British quarterly magazine called "The Investigator" published in 1823

In England, paupers in workhouses were often made to crush bones. In 1845 some of the inmates of the workhouse in Andover were so hungry that they resorted to eating any raw meat that was still present on the bones and this caused a national scandal (Figure c).

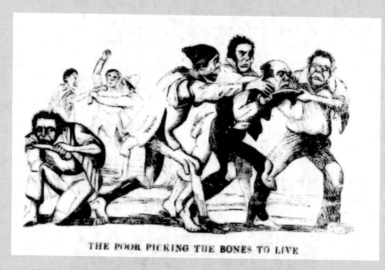

THE POOR PICKING THE BONES TO LIVE

Figure (c) Newspaper illustration from *The Penny Satirist* dated 1845-09-06, titled "The Andover Bastille" and subtitled "The poor picking the bones to live." An illustration to the newspaper's article about the conditions inside the Andover Union workhouse, where starving inmates ate bones meant for use in fertilizer.
Public domain via Wikimedia Commons

Surprisingly, very little is known about the effects that human corpses have on the survival or growth of neighbouring plants. However, there are a number of reports of how plant growth is affected by the decomposition of the corpses of other mammals. Unfortunately, it's difficult to generalise from the results of these studies because which specific plants are affected depends so much on the predominant types that exist in that particular region i.e. forest, grassland, tundra or desert. Nevertheless, it's worth looking at the results of two interesting studies. The first of these was carried out in Holland and involved placing red deer carcasses in grassy regions of a nature reserve in April (Figure 7.13a). By August only dry remains were present and the sites

(a)

(b)

Figure 7.13 Effect of red deer carcases on plant growth in a nature reserve in Holland (a) Red deer carcase deposited on grassland.
(b) The site with the dry remains of a red deer after 5 months. Vigorous plant growth can be seen, this was dominated by the thistle *Carduus crispus*.
van Klink R. *et al.* Rewilding with large herbivores: Positive direct and delayed effects of carrion on plant and arthropod communities. *PLoSONE* 2020; 15(1): e0226946. https://doi.org/10.1371/journal.pone.0226946
This is an open access article distributed under the terms of the Creative Commons Attribution License

were overgrown by plants, the dominant one being the thistle *Carduus crispus* (Figure 7.13b). The plant mass was 5 times greater than at sites that didn't have a carcase, showing the dramatic effect of the additional corpse-derived nutrients. There were also increased quantities of *Sisymbrium officinale* (hedge mustard), *Plantago major* (greater plantain) and various grasses.

Interestingly, the study also showed that the nutritional content of the plants that grew at the corpse site was significantly greater than that of those that grew at carcase-free sites. Other studies have shown that this increased plant growth can persist for more than 10 years.

In another study, carcases of bison, cattle and deer were left on prairie grassland in Kansas and the plants growing within a 50 cm radius of the centre of the carcases were analysed over a 5 year period. A distinct plant succession was apparent over this period and these results are summarised in Figure 7.14.

The first thing to notice is the low abundance of plants one year after placement of the corpses – this is characteristic of a CDI. There were only small quantities of *Ambrosia psilostachya* (perennial ragweed), *Euphorbia glyptosperma* (ridge-seed spurge) and *Andropogon gerardii* (big bluestem). After two years, however, a number of plants were abundant and these included *Ambrosia psilostachya*, *Ambrosia artemisiifolia* (common ragweed) and *Chenopodium berlandieri* (pitseed goosefoot). *Ambrosia psilostachya* was still abundant after 4 years and, at this time point, the site also had high proportions of *Agropyron smithii* (Western wheatgrass) and *Euphorbia glyptosperma*. After 5 years, *Agropyron smithii* dominated the site.

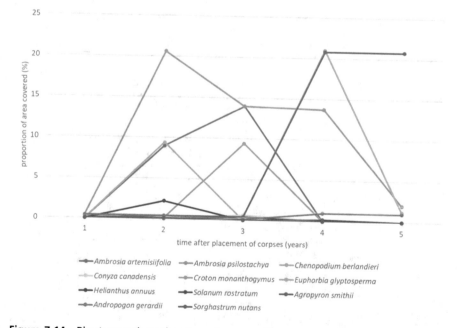

Figure 7.14 Plant growth at the sites of mammalian corpses over a five year period. Plants within a 50 cm radius of the centre of each corpse were identified and quantified.

The general conclusions that can be drawn from these, and other studies, are that mammalian corpses have an initial negative impact on the local vegetation but this is followed by a succession of plants that take advantage of the increased soil fertility. This soil enrichment has a long-term effect on the local vegetation producing not only greater quantity of plants but ones that also have an increased nutritional content.

Box 7.4 Graveyards and yew trees

In most of the graveyards of England, Wales, Scotland, Ireland and Northern France you'll see a yew tree. This association has inspired many poets, among them Alfred Lord Tennyson who, in 1849, wrote the following lines about a yew tree in a graveyard as part of his poem "In Memoriam"

> Old warder of these buried bones,
> And answering now my random stroke
> With fruitful cloud and living smoke,
> Dark yew, that graspest at the stones
> And dippest toward the dreamless head,
> To thee too comes the golden hour
> When flower is feeling after flower;
> But Sorrow--fixt upon the dead,
> And darkening the dark graves of men,--
> What whisper'd from her lying lips?
> Thy gloom is kindled at the tips,
> And passes into gloom again

Some of these yews are very old and have been estimated to be about five thousand years old (Figure). What's the reason for this association of yews with graves? A number of explanations have been put forward, based on folk-lore, so there's no scientific evidence to support any of them (Old men's fancies: the case of the churchyard yew. Chandler J. *Folklore Society News* 15: 3-6; The Ancient Yew. Bevan-Jones R, 2002).

They include:

- They thrive on corpses and so were planted in graveyards to provide a good source of wood for the longbow (a favourite weapon in English armies)
- They absorb the odours produced by corpse putrefaction
- Yew trees for making bows were planted in churchyards where they wouldn't be eaten by, and poison, grazing animals.
- They were planted in churchyards because they are poisonous to animals and so farmers would make sure that their animals didn't stray into them and damage graves.
- When freshly cut, the heartwood is red and the sapwood is white - these colours symbolise the blood and body of Christ
- because they are evergreen, their foliage symbolised the resurrection of the body

Box 2.1 (continued)

Figure. The Crowhurst Yew in England which is estimated to be 4,000 years old. Although the trunk is hollow, the tree remains in good health
Image courtesy of Peter Trimming. Attribution 2.0 Generic (CC BY 2.0) via Flickr

7.1.3.7 The Bigger Picture

Many of the insects that lay their eggs on a corpse will then wander off and feed on nectar and, in doing so, will pollinate these flowers. The eggs they've laid will eventually produce maggots that feed on the corpse, these will then pupate and develop into adults which will themselves then go off to pollinate flowers. The corpse, therefore, acts as a hub for the production of plant-pollinating insects. Although bees are the most important plant-pollinators, flies are the next most important. The insects that behave in this way include blow flies, flesh flies, house flies and *Fannia* species. As well as being able to pollinate a wide variety of plants in general, research has shown that flies are frequent pollinators of a number of plants that are important as food for humans (Table 7.4). So, as well as providing food for insects, your corpse will also be helping to feed humans (Figure 7.15) – isn't that a comforting thought?

However, this represents just one of the many types of "food chain" that we become part of once we die. There are so many of these that they form a huge network of interactions known as a "food web" and a simplified version of this is shown in Figure 7.16

Table 7.4 Flies found on corpses that are also important pollinators of plants used as food by humans

Insect	Plants pollinated
Blow flies	Avocado, blueberries, brussels sprouts, carrot, macadamia, mango, onion, leek, lucerne, pak choi, strawberry, pear, peach, plum
House flies (*Musca* species)	Leek, mango, onion, pak choi, pear, cherry, peach, plum
Flesh flies	Onion, pak choi
Fannia species	Pear, cherry, peach, plum

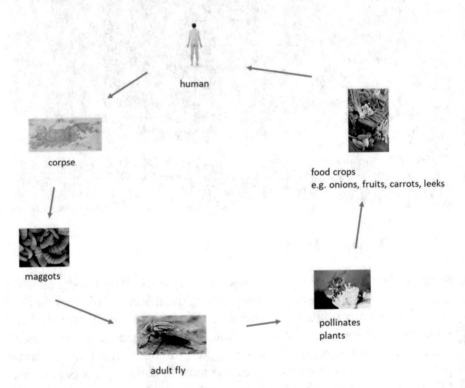

human

corpse

maggots

adult fly

food crops
e.g. onions, fruits, carrots, leeks

pollinates
plants

Figure 7.15 An interesting cycle, food for humans from a human corpse
Vegetables USDA no author noted, Public domain, via Wikimedia Commons
Human: Mikael Häggström, CC0, via Wikimedia Commons
Corpse: Images courtesy of the Collectie Stad Antwerpen, MAS; Public Domain, CC0 1.0 Universal (CC0 1.0),
Maggots: MD-Terraristik – Laut [1] ist Dennis Kress Mitinhaber des Unternehmens, Public domain, via Wikimedia Commons
Fly: Saleem Hameed <saleemham at rediffmail.com>, CC BY 2.5 <https://creativecommons.org/licenses/by/2.5>, via Wikimedia Commons
Flower being pollinated: Image courtesy of Derek Keats, Attribution 2.0 Generic (CC BY 2.0)

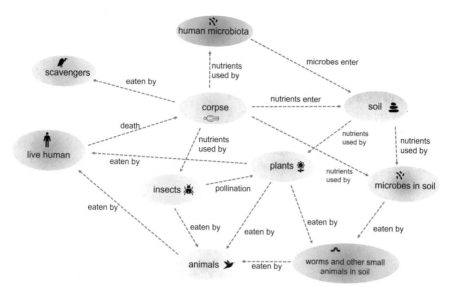

Figure 7.16 Some of the nutrient-recycling interactions in which humans and their corpses are involved

7.2   Final Thoughts

I hope you've enjoyed this description of the fascinating fate that awaits all of us. To summarise, and remind you what it's all about, here's a short poem written especially for you during the COVID-19 pandemic:

Now in lockdown number three
How many more are there to be?
A fine day assured by a bright, clear dawn
Yet my mind is filled with thoughts forlorn.
The grass is green and snowdrops white
My heart should tremble with delight.
But the chemistry of life and death
Hangs heavy on my every breath.
The life-giving soil that nurtures flowers

Must be renewed from flesh like ours.
Nutrients that now through my body course
Will one day be the very source
Of food for grass and sprouting seeds
And will satisfy their growing needs.
So why be glum with eyes downcast
My body will fade but I'll outlast
A thousand billion generations
So this is cause for celebrations.
Recycled will be my flesh and skin
So I'll live on and in the end will win. (Figure 7.17).

Figure 7.17 Snowdrops growing in the graveyard of St Frideswide's Church in the English village of Frilsham. These plants are, no doubt, making use of the nutrients supplied by the many bodies that have been buried there. The church dates from the 12th century and is believed to have been built on the site of an earlier Saxon church

7.3 Want to Know More?

An introduction to ecology
https://www.youtube.com/watch?v=GlnFylwdYH4

Ecology: examining the relationships between living things
https://www.environmentalscience.org/ecology

Soil. An introduction from the UK's Environment Agency
https://assets.publishing.service.gov.uk/government/uploads/system/
uploads/attachment_data/file/805926/State_of_the_environment_soil_
report.pdf

What is Soil? - Definition, Structure & Types
https://study.com/academy/lesson/what-is-soil-definition-structure-
types.html

Mortality data from UN
https://www.un.org/en/development/desa/population/publications/pdf/
mortality/WMR2019/WorldMortality2019DataBooklet.pdf

Rewilding with large herbivores: Positive direct and delayed effects of carrion
on plant and arthropod communities. van Klink R, van Laar-Wiersma J, Vorst
O, Smit C *PLoSONE* 2020; 15(1): e0226946.
https://doi.org/10.1371/journal.pone.0226946

The corpse project. Can we lay bodies to rest so that they help the living and
the earth?
http://www.thecorpseproject.net/

Soil science related to the human body after death
http://www.thecorpseproject.net/wp-content/uploads/2016/06/Corpse-and-
Soils-literature-review-March-2016.pdf

The chemistry decomposition in human corpses. Ioan BG *et al. Revista de
Chimie, Bucharest* 2017; 68(6):1450-1454. DOI: https://doi.org/10.37358/
RC.17.6.5672
https://www.funeralnatural.net/sites/default/files/articulo/archivo/humande-
composition_0.pdf

An evaluation of soil chemistry in human cadaver decomposition islands:
Potential for estimating postmortem interval (PMI). Fanchera JP *et al. Forensic
Science International* 2017; 279, 130-139

Functional and structural succession of soil microbial communities below decomposing human cadavers. Cobaugh KL, Schaeffer SM, DeBruyn JM *PLoS ONE* 2015; 10(6):e0130201
https://www.ncbi.nlm.nih.gov/pmc/articles/PMC4466320/

The role of flies as pollinators of horticultural crops: an Australian case study with worldwide relevance. Cook DF et al. *Insects* 2020, *11*(6), 341; https://www.mdpi.com/2075-4450/11/6/341/htm

Spatial impacts of a multi-individual grave on microbial and microfaunal communities and soil biogeochemistry. Keenan SW *et al. PLoS One* 2018 Dec 12;13(12):e0208845. doi: https://doi.org/10.1371/journal.pone.0208845. https://www.ncbi.nlm.nih.gov/pmc/articles/PMC6291161/

Soil nematode functional diversity, successional patterns, and indicator taxa associated with vertebrate decomposition hotspots. Taylor LS *et al.. PLoS One.* 2020 Nov 4;15(11):e0241777. doi: https://doi.org/10.1371/journal. pone.0241777.
https://www.ncbi.nlm.nih.gov/pmc/articles/PMC7641364/

Cadaver decomposition and soil: processes. Carter DO, Tibbett M. file://C:/Users/mikew/AppData/Local/Temp/CarterTibbett Cadaverdecompsoil-TaphonomyBook2007.pdf

Comparative decomposition of humans and pigs: soil biogeochemistry, microbial activity and metabolomic profiles.
DeBruyn JM *et al. Frontiers in Microbiology* 2021 Jan 13;11:608856. doi: https://doi.org/10.3389/fmicb.2020.608856. eCollection 2020
https://pubmed.ncbi.nlm.nih.gov/33519758/

Temporal and spatial impact of human cadaver decomposition on soil bacterial and arthropod community structure and function. Sing B *et al. Frontiers in Microbiology* 2018 Jan 4;8:2616. doi: https://doi.org/10.3389/fmicb.2017.02616
https://www.ncbi.nlm.nih.gov/pmc/articles/PMC5758501/

Potential groundwater pollutants from cemeteries. The Environment Agency, UK. 2004
https://www.funeralnatural.net/sites/default/files/articulo/archivo/groundwatercontamination.pdf

The impact of cemeteries on the environment and public health. World Health Organisation https://apps.who.int/iris/bitstream/handle/10665/108132/EUR_ICP_EHNA_01_04_01(A).pdf;jsessionid=9297A082F515721283964B45844B1176?sequence=1

Cemeteries and burials: prevent groundwater pollution. Environment Agency, UK, 2017
https://www.gov.uk/guidance/cemeteries-and-burials-prevent-groundwater-pollution

Effects of decomposing cadavers on soil nematode communities over a one-year period. Szelecz I. *et al. Soil Biology & Biochemistry* 2016; 103; 405e416

The role of carrion in maintaining biodiversity and ecological processes in terrestrial ecosystems. Barton PS *et al., Oecologia* 2013;171(4):761-72. doi: https://doi.org/10.1007/s00442-012-2460-3.

Cadaver decomposition in terrestrial ecosystems. Carter DO. *et al., Naturwissenschaften.* 2007;94(1):12-24. doi: https://doi.org/10.1007/s00114-006-0159-1.

Review of human decomposition processes in soil. Dent BB *et al. Environmental Geology* 2004; 45; 576–585

Microbial signatures of cadaver gravesoil during decomposition. Finley SJ *et al. Microbial Ecology.* 2016 Apr;71(3):524-9. doi: https://doi.org/10.1007/s00248-015-0725-1.

Impact of cemeteries on groundwater chemistry: A review. Józef Żychowski. *CATENA* 2012; 93; 29-37

Appendix I. Glossary

Adipocere	Also known as "grave wax". A waxy coating that sometimes forms on the outside of a corpse, particularly when it's immersed in water.
Aerobe	A creature that needs oxygen to live. Human beings are aerobes.
Anaerobe	A creature that can live in the absence of oxygen. Many bacteria can do this but humans and animals can't.
Autolysis	A process involving the self-destruction of a cell. The protective membrane that encloses the cell is disrupted thereby allowing the cell contents to escape.
Bacteraemia	The presence of bacteria in the bloodstream
Bacteriocin	A compound produced by bacteria that can kill or inhibit other bacteria.
Biodegradable	Capable of being decomposed by living organisms.
Biodiversity	A measure of how many different species live in a particular ecosystem, area or region.
Biofilm	An aggregate of large numbers of microbes on a surface that is enclosed in a jelly-like substance (known as a matrix).
Colon (bowel, large intestine)	An organ of the human digestive system. It absorbs water and electrolytes from the contents of the digestive tract as well as the nutrients produced by the enormous number of microbes that live there.
Cytotoxin	A compound that is able to kill one or more types of human cells.

© The Author(s), under exclusive license to Springer Nature Switzerland AG 2022
M. Wilson, *Life After Death: What Happens to Your Body After You Die?*,
Springer Praxis Books, https://doi.org/10.1007/978-3-030-83036-6

Deoxyribonucleic acid (DNA)	A complex molecule that, in most organisms, is the hereditary material containing all the instructions an organism needs to develop, live and reproduce. It's made up of chemical building blocks known as nucleotides. Each nucleotide consists of a sugar (deoxyribose), a phosphate group and one of four types of nitrogen bases – adenine, cytosine, thymine and guanine. The order, or sequence, of these bases determines what biological instructions are contained in a strand of DNA.
Eccrine glands	The glands in the skin that produce sweat
Electron micrograph	A photograph taken through an electron microscope.
Environmental determinant	Any factor of an environment that affects the ability of a microbe to colonise, or grow within, that environment.
Enzyme	A protein that acts as a catalyst i.e. it speeds up a chemical reaction. Some function inside cells (intracellular) while other work outside cells (extracellular).
Facultative anaerobe	A creature that can live in the presence or absence of oxygen. Many bacteria and fungi can do this.
Fimbriae	Hair-like projections on the surface of a microbe that are involved in adhesion of the microbe to surfaces.
Flagellum (plural: flagella)	A whip-like structure that enables a microbe to move
Glycoprotein	A protein that has sugar molecules attached to it.
Glycosaminoglycan	A type of polysaccharide. It contains complex monosaccharides that have carboxylic acid and amino groups attached. They are important components of connective tissue. Examples include heparan sulphate, chondroitin sulphate, keratan sulphate and hyaluronic acid.
Glycosidase	An enzyme that breaks down the bond between sugar molecules in a polysaccharide or oligosaccharide. These large molecules are, therefore, broken down into small sugar molecules
Gram stain	A staining procedure that is very important in identifying bacteria. All bacteria can be described as being either Gram-positive or Gram-negative depending on their appearance after they've undergone the staining procedure.
Habitat	The place where a particular organism lives because the environmental conditions there are suitable for its growth and survival

Host	The larger organism in a symbiotic relationship.
Hydrolase	An enzyme that can break down a macromolecule into smaller molecules. Examples include proteases, glycosidases, lipases and nucleases.
Hydrolysis	Literally "splitting with water". The chemical decomposition of a substance by water. In living organisms this is usually carried out by enzymes
Hydroxyapatite	A mineral (consisting mainly of calcium phosphate and calcium carbonate) found in the bones and teeth of humans and other animals.
Imago	The fully-developed, adult stage of an insect
Keratinocyte	An epidermal cell that produces keratin. It is the predominant cell type found in the epidermis.
Larva (plural larvae)	An early stage in the development of an insect
Lipase	An enzyme that can breakdown a lipid to smaller molecules such as fatty acids and glycerol.
Lipid	A large molecule that consists of one or more fatty acid molecules attached to a glycerol molecule. Lipids are insoluble in water and are important components of cell membranes.
Lymphocyte	A type of white blood cell that is made in the bone marrow and is found in the blood and in lymph tissue. The two main types of lymphocytes are B lymphocytes and T lymphocytes. B lymphocytes make antibodies, and T lymphocytes help kill tumour cells and infected cells and help control immune responses.
Macrophage	A large white blood cell that is involved in the detection, phagocytosis and destruction of bacteria and other microbes.
Macromolecule	Literally, a large molecule. There is no rigid definition of the number of atoms a molecule must contain before it is classed as being a macromolecule, although they generally contain more than 1,000 atoms. Examples of important macromolecules found in living organisms include proteins, polysaccharides, nucleic acids, lipids, glycoproteins and proteoglycans.
Metamorphosis	A dramatic change in shape. Insects usually go through a number of such changes during their life cycle.
Microbiota	(also sometimes referred to as the **Microbiome**) The microbial community present at a particular site or associated with an organ system or entire organism. It includes all of the bacteria, archaea, fungi, viruses, algae and protozoa present in that community.

Motile	Able to move.
Mucinase	An enzyme that can breakdown mucins.
Mucins	Large glycoproteins that are produced by the epithelium and are a major constituent of mucus. Some remain attached to epithelial cells while others are secreted into the mucus layer that covers mucosal surfaces.
Mucus	The watery layer that coats all internal body surfaces that are exposed to the environment. These include the respiratory, gastrointestinal, reproductive and urinary tracts.
Multicellular	An organism that consists of a number of cells. Many fungi and algae are multicellular, whereas bacteria, protozoa and archaea are unicellular, i.e. they consist of a single cell.
Neutrophil (Neutrophilic granulocyteor polymorphonuclear neutrophil)	A type of white blood cell whose main function is to phagocytose and kill microbes. Neutrophils are the most abundant white blood cells
Nomenclature	The system used in the naming of organisms.
Nuclease	An enzyme that can break down nucleic acids to smaller molecules such as nucleotides, nucleosides, sugars etc.
Oligosaccharide	A molecule that consists of between 2 and 9 monosaccharide (sugar) units.
Organelle	A minute structure within a living cell that has a specific function to perform. For example a mitochondrion – its job is to produce energy for the cell.
Pathogen	A microbe that is able to cause disease in a human, other animals or plants.
Peptide	A molecule that is composed of between 2 and 50 amino acids i.e. it is a small protein.
Peptidase	An enzyme that can break down a peptide or polypeptide to amino acids.
pH	A measure of how acidic or alkaline a solution is. The scale is logarithmic and ranges from 0 to 14 with acidic solutions having a pH less than 7 while alkaline ones have a pH greater than 7. Pure water is neutral i.e. it has a pH of 7.
Phagocyte	A cell that can ingest other cells or particles
Photomicrograph	A photograph taken through a microscope.
Physiology	The study of how a living organism functions.

Polypeptide	A molecule that is composed of between 2 and 50 amino acids i.e. it is a small protein.
Polysaccharide	A macromolecule that consists of more than 9 monosaccharide units.
Polysaccharidase	An enzyme that can break down polysaccharides to smaller molecules such as oligosaccharides and sugars.
Protease (or proteinase)	An enzyme that can break down a protein to smaller molecules such as peptides and amino acids.
Proteoglycan	A macromolecule that consists of glycosaminoglycan chains attached to a protein. They are important components of the extracellular matrix.
Pupa	A stage in the development of an insect that occurs between the larval and adult stages.
Ribonucleic acid (RNA)	Ribonucleic acid is a complex molecule similar to DNA. It's made up of chemical building blocks known as nucleotides. Each nucleotide consists of a sugar (ribose), a phosphate group and one of the four types of nitrogen bases – adenine, cytosine, uracil and guanine. Its main role in most organisms is to convert the genetic information encoded by DNA into proteins. However, in some viruses, it functions as the hereditary material, i.e. it replaces DNA as the molecule containing the microbe's genetic instructions.
Sebum	An oily substance produced by the sebaceous glands of the skin.
Spore	In bacteria this is a protective structure that enables the organism to survive harsh environmental conditions such as temperature extremes, dessication etc. Not all bacteria are able to form spores. In fungi, a spore is a reproductive structure. All fungi produce spores.
Stratum corneum	The outer layer of the skin. It consists of dead skin cells known as keratinocytes.
Symbiont	An organism that is very closely associated with another, usually larger, organism.
Symbiosis	The living together of two or more dissimilar organisms.
Toxin	A compound produced by a living organism that can damage or kill other organisms. Some microbes can produce toxins that can harm, or kill, humans.
Unicellular	An organism that consists of just one cell. Bacteria, protozoa and archaea are unicellular. In contrast, most fungi and algae are multicellular.

Virulence factor	Any component of a microbe that enables it to withstand its host's antimicrobial defence systems or that causes damage to its host.
Volatile	A substance that readily turns into a gas at room temperature.

Appendix II. Descriptions of Microbes Mentioned in the Book

Unless otherwise stated, the names refer to genera of microbes. When a microbial genus is listed as being present in humans, very often it will also have been found in other animals. Space restrictions, however, have meant that we have omitted long lists of which other animals also harbour that type of microbe. In the column listing whether or not that type of microbe can break down any of the macromolecules present in human tissues, we have also noted if it produces a toxin that can kill human cells. This is important as such killing will release a large number of valuable nutrients from the cell.

Microbial genus/family/order/phylum	Usual habitat and main characteristics	Ability to break down macromolecules and kill human cells
Acinetobacter	• found in soil, water, humans (mainly on skin) • Gram-negative coccobacilli • aerobes • grow over the temperature range 20 - 42°C	• hydrolyse proteins • hydrolyse lipids • hydrolyse polysaccharides • produce toxins that kill human cells
Actinomyces	• found in soil, humans (mainly in mouth and gut) • Gram-positive filaments • display branching • non-motile • non-sporing • most are facultative anaerobes but some are obligate anaerobes • grow over the range 20 - 45°C	• hydrolyse proteins • hydrolyse lipids • hydrolyse polysaccharides • hydrolyse glycoproteins

(continued)

M. Wilson, *Life After Death: What Happens to Your Body After You Die?*, Springer Praxis Books, https://doi.org/10.1007/978-3-030-83036-6

(continued)

Microbial genus/family/order/phylum	Usual habitat and main characteristics	Ability to break down macromolecules and kill human cells
Akkermansia	• found in humans (mainly in gut) • Gram-negative coccobacilli • non-motile • non-sporing • anaerobes • grow over temperature range 15 - 40°C • grow over pH range 5.0 - 8.0	• hydrolyse mucins • hydrolyse polysaccharides • hydrolyse proteins
Alcaligenes	• found in soil, water, humans (mainly in gut) • Gram-negative bacilli • motile • aerobes	• hydrolyse proteins • hydrolyse lipids
Arthrobacter	• found in soil, humans (mainly in gut) • Gram-positive bacilli • motile • aerobes	• hydrolyse nucleic acids • hydrolyse proteins • hydrolyse polysaccharides • hydrolyse lipids
Aspergillus	• found in soil, humans (mainly in gut) • fungi • aerobes	• hydrolyse polysaccharides • hydrolyse proteins • hydrolyse lipids • hydrolyse nucleic acids
Aureobasidium	• found in soil, water, humans (mainly in mouth) • fungi • aerobes	• hydrolyse polysaccharides • hydrolyse proteins • hydrolyse lipids • hydrolyse nucleic acids • produce toxins that kill human cells
Bacillaceae	• bacterial family • consists of 19 genera • Gram-positive bacilli • most are motile • most form spores • most are aerobes or facultative anaerobes	

(continued)

(continued)

Microbial genus/family/ order/phylum	Usual habitat and main characteristics	Ability to break down macromolecules and kill human cells
Bacillus	• found in soil, water, humans (mainly in gut) • Gram-positive bacilli • motile • form spores • aerobes or facultative anaerobes • grow over temperature range 5 - 55°C	• hydrolyse polysaccharides • hydrolyse proteins • hydrolyse lipids • hydrolyse nucleic acids • hydrolyse chondroitin • produce toxins that kill human cells
Bacteroides	• found in humans (mainly in gut, respiratory tract) • Gram-negative bacilli • most species are non-motile • non-sporing • anaerobes	• hydrolyse polysaccharides • hydrolyse proteoglycans • hydrolyse proteins • hydrolyse mucins • hydrolyse hyaluronan • hydrolyse heparan • hydrolyse chondroitin sulphate
Bacteroidetes	• a phylum of bacteria • Gram-negative bacilli • aerobes or anaerobes • non-sporing • contains three classes	
Bifidobacterium	• found in humans (mainly in gut, mouth) • Gram-positive bacilli • exhibit branching • non-sporing • non-motile • anaerobes • grow over the temperature range 8 - 49.5°C • grow over pH range 4.0 - 8.5	• hydrolyse polysaccharides • hydrolyse mucins • hydrolyse proteins
Blautia	• found in humans (mainly in gut) • Gram-positive bacilli • anaerobes • non-motile	• hydrolyse polysaccharides • hydrolyse proteins • hydrolyse lipids

(continued)

(continued)

Microbial genus/family/order/phylum	Usual habitat and main characteristics	Ability to break down macromolecules and kill human cells
Brevibacterium	• found in soil, humans (mainly on skin) • Gram-positive bacilli • non-motile • aerobes • grow over temperature range 4 - 42°C • grow over pH range 5.5 – 10.0	• hydrolyse proteins • hydrolyse lipids • hydrolyse nucleic acids
Campylobacter	• found in soil, water, humans (mainly in gut, mouth) • Gram-negative, curved bacilli • motile • grow best in low concentrations of oxygen	• hydrolyse proteins • hydrolyse nucleic acids • produce toxins that kill human cells
Candida	• found in soil, humans (mainly in gut, mouth) • dimorphic fungi - exists as oval cells and hyphae • facultative anaerobes • grow best under aerobic conditions • grow over the temperature range 20 - 40°C • aciduric and acidogenic • grow over pH range 2 - 8	• hydrolyse proteins • hydrolyse lipids • hydrolyse polysaccharides • hydrolyse mucins • hydrolyse hyaluronan • produce toxins that kill human cells
Capnocytophaga	• found in humans (mainly in mouth) • Gram-negative bacilli • motile • facultative anaerobes	• hydrolyse proteins • hydrolyse polysaccharides • hydrolyse lipids • hydrolyse mucins
Cladosporium	• found in soil, humans (mainly in gut, mouth) • fungi • aerobes	• hydrolyse proteins • hydrolyse lipids • hydrolyse polysaccharides • produce toxins that kill human cells
Clostridiaceae	• A family of bacteria • Gram-positive bacilli • anaerobes • form spores • usually motile • contains 34 genera	
Clostridiales	• an order of bacteria • contains 16 families	

(continued)

Microbial genus/family/order/phylum	Usual habitat and main characteristics	Ability to break down macromolecules and kill human cells
Clostridium	• found in soil, water, humans (mainly in gut) • Gram-positive bacilli • produce spores • anaerobes • most species are motile • optimum growth of most species occurs between 30°C and 40°C	• hydrolyse polysaccharides • hydrolyse proteins • hydrolyse mucins • hydrolyse lipids • hydrolyse nucleic acids • hydrolyses proteoglycans • hydrolyse hyaluronan • hydrolyse chondroitin • produce toxins that kill human cells
Corynebacterium	• found in soil, water, humans (mainly on skin, in mouth) • Gram-positive bacilli • pleomorphic • non-sporing • non-motile • facultative anaerobes or aerobes • grow over temperature range 15 - 40°C	• hydrolyse lipids • hydrolyse nucleic acids • produce toxins that kill human cells
Cutibacterium	• found in humans (mainly on skin, respiratory tract) • Gram-positive bacilli • pleomorphic • obligate anaerobes or microaerophiles • grow over pH range 4.5 - 8.0 • non-motile • non-sporing	• hydrolyse proteins • hydrolyse lipids • hydrolyse hyaluronan • hydrolyse nucleic acids • hydrolyse chondroitin
Dermabacter	• found in humans (mainly on skin) • Gram-positive bacilli • facultative anaerobes • non-motile	• hydrolyse proteins • hydrolyse polysaccharides • hydrolyse nucleic acids
Entamoeba	• found in soil, water, humans (mainly in gut, mouth) • protozoa • facultative anaerobes	• hydrolyses proteins • hydrolyses nucleic acids • hydrolyse polysaccharides • produce toxins that kill human cells

(continued)

(continued)

Microbial genus/family/ order/phylum	Usual habitat and main characteristics	Ability to break down macromolecules and kill human cells
Enterococcus	• found in soil, water, humans (mainly in gut) • Gram-positive cocci • facultative anaerobes • non-sporing • most species are non-motile • grow over temperature range 5 - 65°C • grow over pH range 4.5 - 10.0	• hydrolyse proteins • hydrolyse polysaccharides • hydrolyse lipids • hydrolyse nucleic acids • hydrolyse hyaluronan • hydrolyse heparan • hydrolyse chondroitin sulphate • produce toxins that kill human cells
Escherichia	• found in humans (mainly in gut) • Gram-negative bacilli • facultative anaerobes • most species are motile • grow over temperature range 15 - 48°C • grow over pH range 5.5 - 8.0	• produce toxins that kill human cells • hydrolyse polysaccharides • hydrolyse proteins • hydrolyse lipids
Enterobacteriaceae	• A family of bacteria • Gram-negative bacilli • facultative anaerobes • usually motile • do not form spores • contains 51 genera	
Eubacterium	• found in humans (mainly in gut, mouth) • Gram-positive bacilli • anaerobes • non-sporing	• hydrolyse proteins • hydrolyse polysaccharides • hydrolyse proteoglycans
Faecalibacterium	• found in humans (mainly in gut) • Gram-negative bacilli • anaerobes • non-sporing • non-motile • grow over pH range 5.0 - 6.7	• hydrolyse proteins • hydrolyse nucleic acids • hydrolyse polysaccharides
Firmicutes	• a phylum of bacteria • most are Gram-positive • cocci or bacilli • many are spore-formers • contains five classes	

(continued)

(continued)

Microbial genus/family/order/phylum	Usual habitat and main characteristics	Ability to break down macromolecules and kill human cells
Fusarium	• found in soil, water, humans (mainly in gut) • fungi • aerobes	• hydrolyse proteins • hydrolyse lipids • hydrolyse polysaccharides • produce toxins that kill human cells
Fusobacterium	• found in humans (mainly in gut, mouth) • Gram-negative bacilli • long, spindle-shaped cells • anaerobes • non-sporing • most species are non-motile	• hydrolyse proteins • hydrolyses proteoglycans • produce toxins that kill human cells
Geotrichum	• found in soil, water, humans (mainly in gut) • fungi • aerobes	• hydrolyse proteins • hydrolyse lipids • hydrolyse polysaccharides
Haemophilus	• found in humans (mainly in mouth, respiratory tract) • Gram-negative pleomorphic bacilli • facultative anaerobes • non-motile • non-sporing • grow over temperature range 20 - 40°C • optimum pH for growth is 7.6	• hydrolyse proteins • hydrolyse nucleic acids • hydrolyse polysaccharides • produce toxins that kill human cells
Ignatzschineria	• found in Sarcophagidae flies • Gram-negative bacilli • aerobes • non-motile • non-sporing	• hydrolyse lipids • hydrolyse proteins
Lactobacillus	• found in soil, water, humans (mainly in gut, mouth, vagina) • Gram-positive bacilli • obligate anaerobes • non-motile • non-sporing • grow over pH range 3.5 - 6.8 • grow over temperature range 15 - 45°C	• hydrolyse polysaccharides • hydrolyse proteins • hydrolyse nucleic acids • hydrolyse lipids • hydrolyse heparan • hydrolyse chondroitin sulphate

(continued)

(continued)

Microbial genus/family/ order/phylum	Usual habitat and main characteristics	Ability to break down macromolecules and kill human cells
Leptotrichia	• found in humans (mainly in mouth) • Gram-negative bacilli • anaerobes • non-motile • non-sporing	• hydrolyse proteins • hydrolyse polysaccharides
Malassezia	• found in humans (mainly on skin) • fungi • exist as oval cells and hyphae • facultative anaerobes	• hydrolyse lipids • hydrolyse proteins
Micrococcus	• found in soil, water, humans (mainly on skin, in gut, in respiratory tract) • Gram-positive cocci, usually in tetrads or clusters • aerobes • non-motile • non-sporing • grow over temperature range 25 - 37°C	• hydrolyse hyaluronan • hydrolyse proteins • hydrolyse nucleic acids • hydrolyse lipids • hydrolyse polysaccharides • hydrolyse mucins
Moraxella	• found in humans (mainly in respiratory tract) • Gram-negative cocci • aerobes • non-motile • non-sporing	• hydrolyse lipids • hydrolyse nucleic acids
Mucor	• found in soil, water, humans (mainly in gut) • fungi • facultative anaerobes	• hydrolyse proteins • hydrolyse lipids • hydrolyse polysaccharides
Mycoplasma	• found in soil, water, humans (mainly in mouth, respiratory tract) • pleomorphic • do not have a cell wall • non-sporing • aerobes or facultative anaerobes • grow over temperature range 20 - 45°C	• hydrolyse proteins • hydrolyse lipids • hydrolyse nucleic acids • hydrolyse mucins • produce toxins that kill human cells

(continued)

(continued)

Microbial genus/family/ order/phylum	Usual habitat and main characteristics	Ability to break down macromolecules and kill human cells
Neisseria	• found in humans (mainly in mouth, respiratory tract) • Gram-negative cocci • cells are often kidney-shaped and in pairs • aerobes • non-sporing • non-motile • grow over temperature range 22 - 40°C • grow over pH range 6.0 - 8.0	• hydrolyse proteins • hydrolyse nucleic acids • produce toxins that kill human cells
Parabacteroides	• found in humans (mainly in gut) • Gram-negative bacilli • anaerobes • non-motile • non-sporing	• hydrolyse proteins
Penicillium	• found in soil, water, humans (mainly in gut) • fungi • facultative anaerobes • grow over temperature range 5 - 37°C	• hydrolyse proteins • hydrolyse polysaccharides • hydrolyse lipids • hydrolyse nucleic acids • hydrolyse mucins
Peptostreptococcus	• found in humans (mainly in gut, vagina, skin, respiratory tract) • Gram-positive cocci • obligate anaerobes • non-motile • non-sporing • optimum growth at 37°C	• hydrolyse proteins • hydrolyse nucleic acids • hydrolyse proteoglycans • hydrolyse hyaluronan • hydrolyse chondroitin
Phascolarctobacterium	• found in humans (mainly in gut) • Gram-negative bacilli • non-motile • non-sporing • anaerobes	
Planococcaceae	• A family of bacteria • Gram-positive • contains 14 genera	

(continued)

(continued)

Microbial genus/family/ order/phylum	Usual habitat and main characteristics	Ability to break down macromolecules and kill human cells
Porphyromonas	• found in humans (mainly in gut, mouth, respiratory tract, vagina) • Gram-negative bacilli • non-motile • non-sporing • obligate anaerobes • optimum growth at 37°C • optimum pH for growth is 7.5	• hydrolyse proteins • hydrolyse polysaccharides • hydrolyse mucins • produce toxins that kill human cells
Prevotella	• found in humans (mainly in gut, mouth, respiratory tract, vagina) • Gram-negative bacilli • non-sporing • non-motile • anaerobes • grow over temperature range 25 - 42°C	• hydrolyse polysaccharides • hydrolyse proteins • hydrolyse nucleic acids • hydrolyse lipids • hydrolyse mucins • hydrolyse hyaluronan • produce toxins that kill human cells
Propionibacterium	• See Cutibacterium	
Pseudomonas	• found in soil, water, humans (mainly in gut) • Gram-negative bacilli • aerobes • motile • non-sporing • grow over temperature range 4 - 42°C	• hydrolyse polysaccharides • hydrolyse proteins • hydrolyse nucleic acids • hydrolyse lipids • hydrolyse hyaluronan • hydrolyse mucins • produce toxins that kill human cells
Roseburia	• found in humans (mainly in gut) • Gram-positive, curved bacilli • anaerobes • motile • non-sporing	• hydrolyse polysaccharides • hydrolyse lipids
Ruminococcus	• found in humans (mainly in gut) • Gram-positive cocci in pairs and chains • anaerobes • some species are motile	• hydrolyse polysaccharides • hydrolyse mucins
Saccharomycetales	• an order of fungi • yeasts that reproduce by budding • contains 13 families	•

(continued)

(continued)

Microbial genus/family/ order/phylum	Usual habitat and main characteristics	Ability to break down macromolecules and kill human cells
Staphylococcus	• found in humans (mainly on skin, in respiratory tract, gut) • Gram-positive cocci, often in clusters • non-motile • non-sporing • facultative anaerobes • grow over temperature range 10 - 45°C • grow over pH range 4.0 - 9.0	• hydrolyse hyaluronan • hydrolyse proteins • hydrolyse polysaccharides • hydrolyse lipids • hydrolyse nucleic acids • produce toxins that kill human cells
Streptococcus	• found in humans (mainly in mouth, respiratory tract, gut) • Gram-positive cocci, usually in pairs or chains • facultative anaerobes • non-sporing • non-motile • grow over temperature range 20 - 42°C	• hydrolyse proteins • hydrolyse polysaccharides • hydrolyse lipids • hydrolyse nucleic acids • hydrolyse mucins • hydrolyse hyaluronan • hydrolyse chondroitin • produce toxins that kill human cells
Streptomyces	• found in soil • Gram-positive filaments • aerobes • grow over pH range 4.0 - 11.5	• hydrolyse polysaccharides • hydrolyse lipids • hydrolyse nucleic acids • hydrolyse proteins • hydrolyse mucins • hydrolyse hyaluronan • hydrolyse chondroitin • produce toxins that kill human cells
Tannerella	• found in humans (mainly in mouth) • Gram-negative bacilli • anaerobes • non-motile • non-sporing	• hydrolyse proteins • hydrolyse polysaccharides • hydrolyse mucins
Tissierellaceae	• family of bacteria • contains 9 genera • Gram-positive bacilli • anaerobes	

(continued)

(continued)

Microbial genus/family/ order/phylum	Usual habitat and main characteristics	Ability to break down macromolecules and kill human cells
Treponema	• found in humans (mainly in mouth, gut) • Gram-negative, spiral-shaped • prefers low oxygen concentrations • motile • non-sporing	• hydrolyse hyaluronan • hydrolyse mucins • hydrolyse proteins
Vagococcus	• found in soil, water, humans (mainly in gut) • Gram-positive cocci • facultative anaerobes • motile • grow over temperature range 10 - 45°C	• hydrolyse proteins • hydrolyse polysaccharides • hydrolyse lipids • produce toxins that kill human cells
Veillonella	• found in humans (mainly in mouth, respiratory tract) • Gram-negative cocci • anaerobes • non-motile • non-sporing • grow over temperature range 24 - 40°C	• hydrolyse proteins • hydrolyse polysaccharides
Wohlfahrtiimonas	• found in Sarcophagidae flies • Gram-negative bacilli • facultative anaerobes • motile • non-sporing	• hydrolyse polysaccharides

Appendix III. Images of Corpses and Other Items That Some Readers May Consider to Be Unpleasant or Disturbing

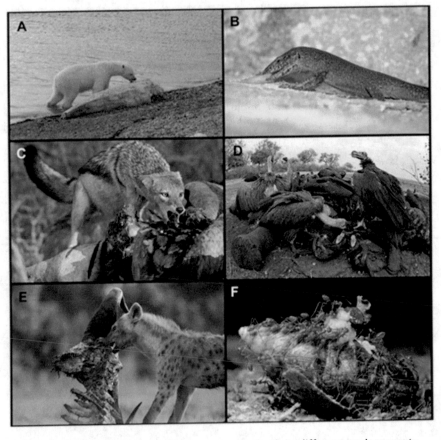

Figure 1.3 Examples of scavenging by various animals in different environments Fallows C, Gallagher AJ, Hammerschlag N (2013) / CC BY (https://creativecommons.org/licenses/by/2.5)

Figure 4.2 The corpse of a pig in the fresh stage of decomposition
Enhanced characterization of the smell of death by comprehensive two-dimensional gas chromatography-time-of-flight mass spectrometry (GCxGC-TOFMS). Dekeirsschieter J et al., *PLoS One.* 2012;7(6):e39005. doi: 10.1371/journal.pone.0039005. Epub 2012 Jun 18. This is an open-access article distributed under the terms of the Creative Commons Attribution License, which permits unrestricted use, distribution, and reproduction in any medium, provided the original author and source are credited.

Figure 4.3 Rigor mortis is apparent in the legs of this dead cow lying beside a road. Spider.Dog from United Kingdom, CC BY 2.0 <https://creativecommons.org/licenses/by/2.0>, via Wikimedia Commons

Figure 4.5 A pig corpse in the bloat stage of decomposition.
Enhanced characterization of the smell of death by comprehensive two-dimensional gas chromatography-time-of-flight mass spectrometry (GCxGC-TOFMS). Dekeirsschieter J et al., *PLoS One*. 2012;7(6):e39005. doi: 10.1371/journal.pone.0039005. Epub 2012 Jun 18. This is an open-access article distributed under the terms of the Creative Commons Attribution License, which permits unrestricted use, distribution, and reproduction in any medium, provided the original author and source are credited.

Figure 4.6 Marbling of the skin in a corpse.
Image from: Pinheiro J. (2006) Decay Process of a Cadaver. In: Schmitt A., Cunha E., Pinheiro J. (eds) Forensic Anthropology and Medicine. Humana Press. https://doi. org/10.1007/978-1-59745-099-7_5

Figure 4.7 Maggots on the corpse of a pig

Figure 4.8 A pig corpse in the active decay stage of decomposition.

Figure 4.9 A pig corpse in the advanced decay stage of decomposition.
Enhanced characterization of the smell of death by comprehensive two-dimensional gas chromatography-time-of-flight mass spectrometry (GCxGC-TOFMS). Dekeirsschieter J et al., *PLoS One*. 2012;7(6):e39005. doi: 10.1371/journal.pone.0039005. Epub 2012 Jun 18. This is an open-access article distributed under the terms of the Creative Commons Attribution License, which permits unrestricted use, distribution, and reproduction in any medium, provided the original author and source are credited.

Figure 4.10 The remains of a pig in the dry state of decay. Bones and hair are clearly visible.
Hbreton19. Permission is granted to copy, distribute and/or modify this document under the terms of the GNU Free Documentation License, Version 1.2 or any later version published by the Free Software Foundation; with no Invariant Sections, no Front-Cover Texts, and no Back-Cover Texts. A copy of the license is included in the section entitled GNU Free Documentation License.

Figure 4.15 A naturally preserved Peruvian mummified male, possibly from the North coast of Peru where the Chimu culture buried their dead in 'mummy bundles', curled up in foetal position with bound hands and feet.
Credit: Wellcome Collection. Attribution 4.0 International (CC BY 4.0)

Box 4.6 The unwrapped mummy of Ramses II who was one of the greatest of the Egyptian pharaohs. He died in 1213 BCE.
Unwrapped Mummy of Ramses II. Credit: Wellcome Collection. Attribution 4.0 International (CC BY 4.0)

Box 6.4 *Dermestes* beetles cleaning a skull
Sklmsta, CC0, via Wikimedia Commons

Box 7.1 White-backed vultures and marabou storks on an elephant carcass in Botswana.
Image courtesy of Michael Jansen via flikr (CC BY-ND 2.0)

Figure 7.5 A cadaver decomposition island visible around a dead pig. The white bar represents 1 metre.

Image from: Cadaver decomposition in terrestrial ecosystems. Carter DO *et al. Naturwissenschaften* 94, 12–24 (2007). https://doi.org/10.1007/s00114-006-0159-1

Index

Printed in the United States
by Baker & Taylor Publisher Services